International Fusion Energy Technology Topical

Topical Conference at the 2008 AIChE Annual Meeting

Philadelphia, Pennsylvania
16 - 21 November 2008

ISBN: 978-1-61567-212-7

Printed from e-media with permission by:

Curran Associates, Inc.
57 Morehouse Lane
Red Hook, NY 12571

Printed by Curran Associates, Inc. (2009)

For permission requests, please contact AIChE
at the address below.

AIChE
3 Park Avenue
New York, NY 10016-5991

Phone: (203) 702-7660
Fax: (203) 775-5177

www.aiche.org

Additional copies of this publication are available from:

Curran Associates, Inc.
57 Morehouse Lane
Red Hook, NY 12571 USA
Phone: 845-758-0400
Fax: 845-758-2634
Email: curran@proceedings.com
Web: www.proceedings.com

TABLE OF CONTENTS

36a Savannah River National Laboratory and the Iter Project

SRNL Director, Savannah River National Laboratory, Aiken, SC 29808

Savannah River National Laboratory, the applied research and development lab at DOE's Savannah River Site, "puts science to work" to create and deploy practical, high-value, cost-effective technology solutions. SRNL has responsibility for the design, fabrication, assembly, testing, and shipment of the Exhaust Processing System for the Iter international fusion energy project. This presentation will detail the laboratory's specific roles in the Iter project.

36b Support for Fusion's Future

Benjamin J. Cross, Energy Security, Savannah River National Laboratory, Aiken, SC 29808

Fusion, the power that runs the sun and the stars, is a viable energy alternative for the future. To make it work, large, unprecedented volumes of hydrogen isotopes must be processed. To facilitate environmental responsibility and economic practicality of fusion technology, the re-use of hydrogen isotopes is vital. A system to recycle the machine exhaust gases into a feed stream containing only hydrogen isotopes is critical.

36c High Average Power Laser Direct Drive Inertial Fusion Energy

Charles A. Gentile, Plasma Physics Laboratory, Princeton University, Princeton, NJ 08543

The United States High Average Power Laser (HAPL) program is developing the conceptual design for a direct drive inertial fusion energy power reactor. This R&D program is a multi-laboratory effort which includes national labs, universities and industry. The reactor, which is planned to be nominally operated at 5 Hz, introduces a D-T cryogenic target into a reactor chamber where it will be symmetrically illuminated by laser light, thus compressing the target and producing a thermonuclear burn. Post detonation energy in the form of neutrons and energetic ions will be harnessed for the end purpose of generating electrical energy. The reactor is fitted with breeder blankets as to produce tritium for use in the device as nuclear fuel. The paper will discuss the status of the HAPL program and the technical challenges associated with the advancement of direct drive IFE power reactor engineering.

36d Technical Challenges of Fusion Energy Development

Samit K. Bhattacharyya, Savannah River National Laboratory, Aiken, SC 29808 and Benjamin J. Cross, Energy Security, Savannah River National Laboratory, Aiken, SC 29808.

Fusion is the energy source of the sun and the stars. On earth, fusion research is aimed at demonstrating that this energy source can be used to produce electricity and/or high grade process heat in a safe and environmentally benign way, with abundant fuel resources, to meet the needs of a growing world population. The development of fusion energy is considered one of the mankind's greatest challenges. This presentation will provide an overview of some of the technical challenges that must be overcome in order for fusion to be an reliable energy source.

88a Predictions of H Isotope Separation Using Crystalline and Amorphous Membranes: A Computational Approach

Lymarie Semidey-Flecha[1], Shiqiang Hao[2], Chen Ling[1], and David S. Sholl[1]. (1) Georgia Institute of Technology, School of Chemical & Biomolecular Engineering, 311 Ferst Drive, N.W., Atlanta, GA 30332-0100, (2) Carnegie Mellon University, Department of Chemical Engineering, Carnegie Mellon University, Pittsburgh, PA 15213

First-principles calculations offer a useful complement to experimental approaches for characterizing H2, D2 and T2 permeance through dense metal membranes as well as amorphous metal membranes. In this study we performed calculations that combine ab initio Density Functional Theory calculations and statistical mechanics to describe the properties of interstitial H and its isotopes in crystalline Pd-based binary and ternary metal alloys, crystalline Fe3B and amorphous Fe3B. This approach predicts the correct isotopic properties of interstitial H in pure Pd, and we have validated it in the past for the membrane properties for a variety of alloy membranes. We used Sieverts' law to calculate solubility of each component in the Pd-based metal membranes and GCMC for the components in Fe3B membranes. A Kinetic Monte Carlo (KMC) scheme was implemented to find diffusion of H, D, and T at a temperature range of $400 \leq T \leq 1200$ K. From these results we are able to predict the permeability of H2, D2 and T2 in each of the membranes and predict which material would be best suited for isotope separation.

88b Iter Tokamak Exhaust Processing System

Bernice Rogers and **Donald J. Green**. Energy Security, Savannah River National Laboratory, Center for Hydrogen Research 999-2W, 773-41A Savannah River Site, Aiken, SC 29808

The ITER project calls for an unprecedented amount of hydrogen isotopes to be processed. To facilitate environmental responsibility and economic application of fusion technology, the re-use of hydrogen isotopes is vital. The US ITER Project Office (USIPO) is responsible for the front end of the ITER Tritium Plant, the Tokamak Exhaust Processing (TEP) System. The TEP system must separate the Tokamak exhaust gases into a stream containing only hydrogen isotopes and a stream containing only non-hydrogen gases. The US's implementation of the TEP system will provide a technically mature, robust, and cost-effective solution for the separation of hydrogen isotopes from the Tokamak exhaust stream.

The USIPO has selected the Savannah River National Laboratory (SRNL) in partnership with the Los Alamos National Laboratory (LANL) to complete the TEP portion of the project. The TEP system consists of a series of filters, catalysts and permeators to achieve a timely separation of hydrogen isotopes from all other species contained in the Tokamak exhaust. The hydrogen isotopes are then sent to a system, furnished by the EU, to deliver deuterium and Tritium for continuous re-injection into the Tokamak. The TEP system must therefore be rugged, maintainable, and have enough surge capacity to meet the anticipated life of 20 years.

88c Iter Pellet Injection Fueling System

Larry R. Baylor[1], Steven J. Meitner[1], and David A. Rasmussen[2]. (1) Oak Ridge National Laboratory (ORNL), Box 2008, Oak Ridge, TN 37831-6169, (2) US ITER WBS Manager Fueling, Oak Ridge National Laboratory (ORNL), Box 2008, Oak Ridge, TN 37831-6169

The fueling system for ITER is designed to provide gas injection to maintain the plasma edge and divertor conditions and solid pellet injection for deep fueling of the reacting plasma. Two pellet injectors are to be provided by the US for fueling using deuterium-tritium pellets with up to 90% tritium content. The pellet fueling rate is specified to be a maximum of 100 Pa-m3/s (50 Pa-m3/s of T2). This is equivalent to 0.23 g/s or ~1.0 cm3/s solid DT to be delivered to the plasma, which is a significant extrapolation from present fusion fueling devices. Screw type extruders are under development to supply this quantity of solid DT for long durations. A repeating gas gun accelerator is planned for the pellet injectors. The screw extruder supplies the DT ice that is cut into pellets and placed in a barrel for acceleration by a low gas throughput propellant gas valve. The injector utilizes a recirculating propellant gas system and a recirculation of the excess extrusion material in order to offload the handling of these gas streams by the tritium plant. Details of the design and results of a prototype twin-screw deuterium extruder experiment will be presented.

88d Iter Tokamak Exhaust Processing Testing

Bryan Carlson, Los Alamos National Laboratory, Los Alamos, NM 87545 and R. Scott Willms, Group Leader, Los Alamos National Laboratory, Los Alamos, NM 87545.

Unprecedented flows and quantities of hydrogen isotopes will be processed by the ITER facility. Catalysts characteristics are required to determine the best catalyst for the ITER application. Research, that has been performed on candidate catalysts, will be presented.

88e Measurement of the ^{3}He Permeability of Deuterium-Tritium-Filled Fused Silica Inertial Confinement Fusion (ICF) Targets to Study the Effects of ^{3}He on Neutron Emission during Implosion

Stephen N. Paglieri[1], Joseph R. Wermer[2], Hailey M. Murdock[2], Arthur Nobile, Jr.[2], Hans W. Herrmann[3], Thomas J. Venhaus[2], James R. Langenbrunner[3], and Joseph M. Mack[3]. (1) Tritium Science and Engineering, Los Alamos National Laboratory, P.O. Box 1663, MS-C927, Los Alamos, NM 87545, (2) Tritium Science & Engineering, Los Alamos National Laboratory, P.O. Box 1663, MS-C927, Los Alamos, NM 87545, (3) Los Alamos National Laboratory, P.O. Box 1663, Los Alamos, NM 87545

A set of laser implosion experiments were conducted at the OMEGA laser at the University of Rochester, Laboratory for Laser Energetics (LLE) to study the effect of ^{3}He concentration in DT-filled target shells on fusion yield in ICF implosions. Eleven laser fusion shells consisting of 1100-μm diameter, hollow, fused silica spheres with 4.6 to 4.7-μm-thick walls were loaded with 520 kPa of deuterium-tritium (DT) and then with ^{3}He (101.3 or 520 kPa). The ^{3}He permeabilities of the shells were determined by measuring the pressure rate of rise into a system with known volume. A mathematical method was developed that relied on the experimental fill pressure and time, and the rate of rise data to solve differential equations using MathCAD to simultaneously calculate ^{3}He permeability and initial ^{3}He partial pressure inside the shell. Because of the high permeation rate for ^{3}He out of the shells compared to that for DT gas, shells had to be kept under ^{3}He until immediately before being laser imploded or "shot" at LLE. The ^{3}He partial pressure in each individual shell at shot time was calculated from the measured ^{3}He permeability. Neutron and gamma yields from an implosion were shown to decrease with increasing ^{3}He addition.

150a Technical Challenges of Fusion Energy

Samit K. Bhattacharyya, Savannah River National Laboratory, Aiken, SC 29808 and Benjamin J. Cross, Energy Security, Savannah River National Laboratory, Aiken, SC 29808.

Fusion is the energy source of the sun and the stars. On earth, fusion research is aimed at demonstrating that this energy source can be used to produce electricity and/or high grade process heat in a safe and environmentally benign way, with abundant fuel resources, to meet the needs of a growing world population. The development of fusion energy is considered one of the mankind's greatest challenges. This presentation will provide an overview of some of the technical challenges that must be overcome in order for fusion to be an reliable energy source.

The Heat Rejection System of the ITER Reactor

G. Dell'Orco, W. Curd, D. Gupta, L. Fan, K.P. Chang, I. Kuehn, A. Kumar*

International Thermonuclear Experimental Reactor Organization (ITER-IO)
13108 Saint Paul lez Durance, France Tel:+33-(0)4-42-25-77-60,Fax:+33-(0)4-42-25-26-00
** ITER INDIA – Bhat, Gandhinagar, Gujarat, 382-428, India*

Abstract

ITER will be the largest fusion experimental reactor in the world designed to reach the first plasma at mid of 2018. ITER is being designed to demonstrate the technological feasibility of the nuclear fusion energy conversion, at plant scale, from high temperature Deuterium-Tritium plasma using the TOKAMAK magnetic confinement arrangement and with a power amplification at least of 10. The ITER design experience, the effort in developing the leading-edge technologies and the knowhow acquired during its operation will guide the realization a proper long-term R&D fusion programme. ITER will bridge the gap between Nuclear Fusion and the large scale commercial production of electricity at competitive cost with other sources by 2045.

ITER will have a large Vacuum Vessel that hosts the Plasma facing components. These components include the Blanket and the Divertor that will operate at temperatures, heat loads and neutron flux higher than that of reached in a nuclear fission power plant reactor.

Among others, one of the main critical issues of the ITER reactor is the transfer of the heat generated in the Plasma, during the D-T reaction, through the Tokamak Cooling Water System (TCWS), to the intermediated closed loop Component Cooling Water System (CCWS) and then, via the open loop Heat Rejection System (HRS) to the environment. The HRS also absorbs heat through the CCWS from other non nuclear systems like the Chilled Water System (CHWS), the Cryogenic System, the Steady State Electrical Power Network (SSEPN) and other auxiliary systems. The HRS rejects to the environment all the heats from the ITER components (nuclear and non nuclear) with the only exception the Vacuum Vessel whose heat are released via a separate primary heat transfer system to the air coolers. The HRS is based on a Cooling Towers System (CTS) designed to release to the environment up to

450 MW whilst the total peak value produced during the DT pulse is about 1270 MW (35% of duty cycle).The CTS requires both high make up water and blowdown flow rates, which in turn demand stringent water chemistry control to limit the possible loop contamination.

This paper describes the main design modifications and optimizations recently introduced with a closed intermediate CCWS and an open HRS system.

The main benefit are the minimization of the capital and operating costs, the optimization of the piping layout, the introduction of a second containment barrier to the gaseous and liquid radwaste releases and the relevant minimization of the impact to the environment.

The Heat Rejection System (HRS)

The HRS provides the final heat sink and rejects all heat loads in the ITER plant with the exception of the Vacuum Vessel.

The main HRS clients are, Fig 1: i) CCWS/1 or the secondary circuits of the main Primary heat Transfer Systems (PHTS) of the TCWS, ii) CCWS/2 or the secondary circuits of the CHWS, the Cryogenic System, the Steady State Electrical Power Network (SSEPN) and other auxiliary systems.

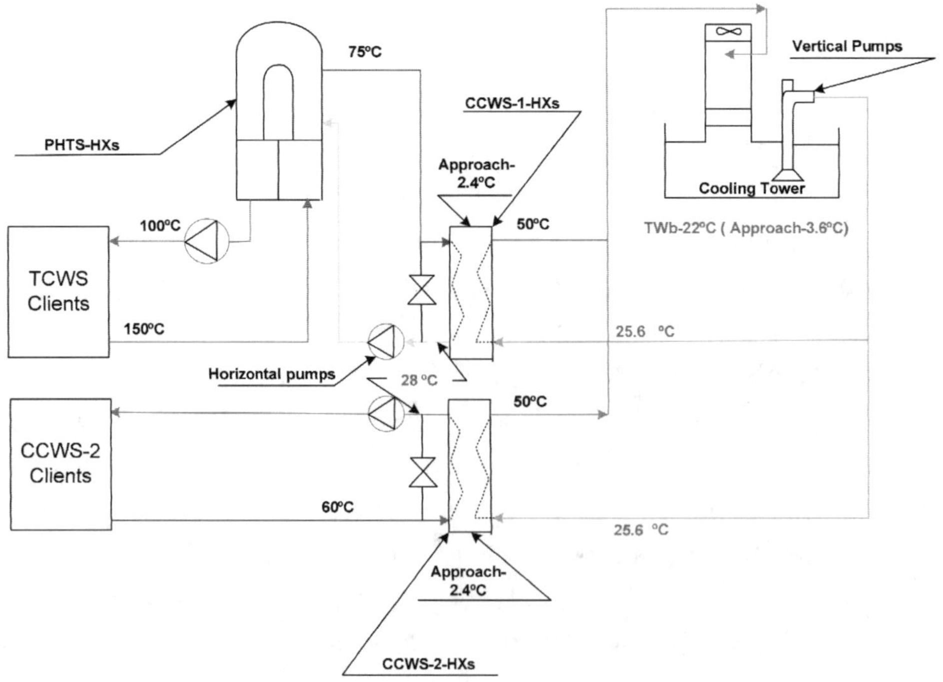

Fig. 1 – ITER Cooling Water System

The introduction of this further CCWS/1 closed loop permits the control of the possible releases of gaseous and solid nuclear products from the TCWS to the HRS. Therefore, the secondary water loop is daily controlled to prevent any normal and abnormal release of contaminants to the HRS.

The HRS consists of the Cooling Tower System (CTS), with 4 air forced/induced draft main units, the water basins, the Vertical Pumps together with the associated water make up and blowdown systems. Whilst the CTS transfers the heat from the CCWS to the atmosphere, the water basin and its ancillary circuits have the following objectives:

1) Provides buffer heat capacity by proper basic water storage to average the heat load during pulsed operation.
2) Provides makeup and blowdown flows to perform the chemical control of the salt concentration in the water basin and to prevent the growth of biologic species (i.e. algae).

The HRS, Fig. 1, provides supply water for the CCWS clients at a maximum temperature of 28 °C and accepts return flow at a maximum temperature of 50°C.

The heat load during the pulse operation are 1270 MW of which 1030 MW in the CCWS/1 from the main PHTS of the TCWS (715 MW from FW/BLK PHTS, 200 MW from DIV/LIM PHTS, and 100 MW from NB injector PHTS), 190 MW from the other CCWS/2 and other 50 MW from the CHWS. The heat load during the dwell condition is 150 MW. The main contributors to the heat load during the dwell condition are the CHWS and the CCWS.

Cooling Tower System (CTS)

The CTS thermal design parameters and features are respectively reported in the Table I and II, whilst the configuration of the wet type CTS along with its bottom basin is shown in Fig. 2. Based on ITER site conditions at Cadarache (France), the wet type CTS consists of 4 cooling cells arranged in an array of tower structures. The overall size of the complete cooling tower envelope is about 62 m (L) x 38 m (W) x 20 m (H), considering two rows of 8 cells (two cells in one unit). The design of the CTS, including the basin, is based on a cost minimisation between cooling tower heat rejection capacity and temperature levelling capability in the basin.

The basin volume is designated as 20.000 m^3, almost equivalent to the circulation volume in one pulse repetition duration (1,800 s) of the reference plasma. This volume allows about 10 hr of pulsed campaign without the water supply. The CTS wet bulb design temperature has been assumed at 22 °C at ITER site at Cadarache. Therefore, in summer (for 45 day from mid July to end August) there is the possibility of CTS operation during the day above the design limit when the wet bulb could statistically reach a peak of 33 °C. In that case, the basin volume provide proper thermal capacity to permit daylight operation (10 h) at the reference design with two silent shifts operation during the night (14 h) to reduce the temperature in the basin. The impact of this possible temporary limitation has been considered negligible on the general machine availability. Both the four units of the cooling tower and the water basin design data have also been positively checked to maintain the cooling requirement during the ITER Non-Inductive Operation I with the pulse duration of 3,000 s at 350 MW of fusion power. The four CTS vertical pumps have 4200 m3/h and 60 mH2O of head capacity, Fig. 3.

Operation Condition	**Reference Plasma**		**Non-Inductive**
Parameter	**Pulse**	**Dwell**	**Burn**
Peak heat duty (MW)	1.270	150	900
Duration (s)	400	1.400	3.000
Flow rate (kg/s)	11.620		
Peak inlet temperature (C)	50	38	55
Max. outlet temperature (C)	25.6	25.6	25.6
Wet bulb temperature (C)	22	22	22
Water basin volume (m^3)	20,000		

Table I - CTS Thermal Design Parameters

Parameter	**Specification**
Number of towers	4
Cells per tower	2
Fans per unit	1
Fan motor (kW)	300
Vertical pump flow rate (m3/h)	4200
Vertical pump head (mH2O)	60

Table II - CTS Design Features

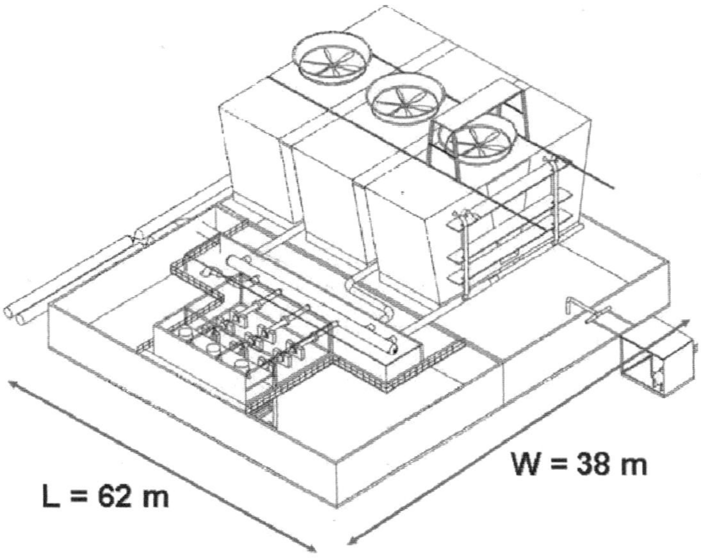

Fig. 2 - Cooling Tower Systems

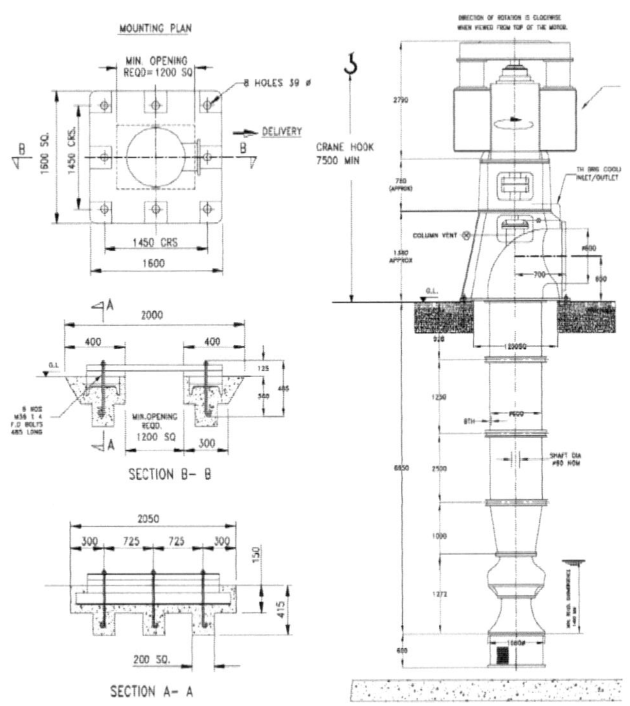

Approximate Weight of the pumpset – 15,000 kg
Above dimensions are tentative and for reference only.

Fig. 3 – CTS Vertical pump

Layout of the HRS

The layout of the HRS is shown in Fig. 4. The pump station and the heat exchangers of the CCWS 1/2 are located on the south side of the CTS and water basin.

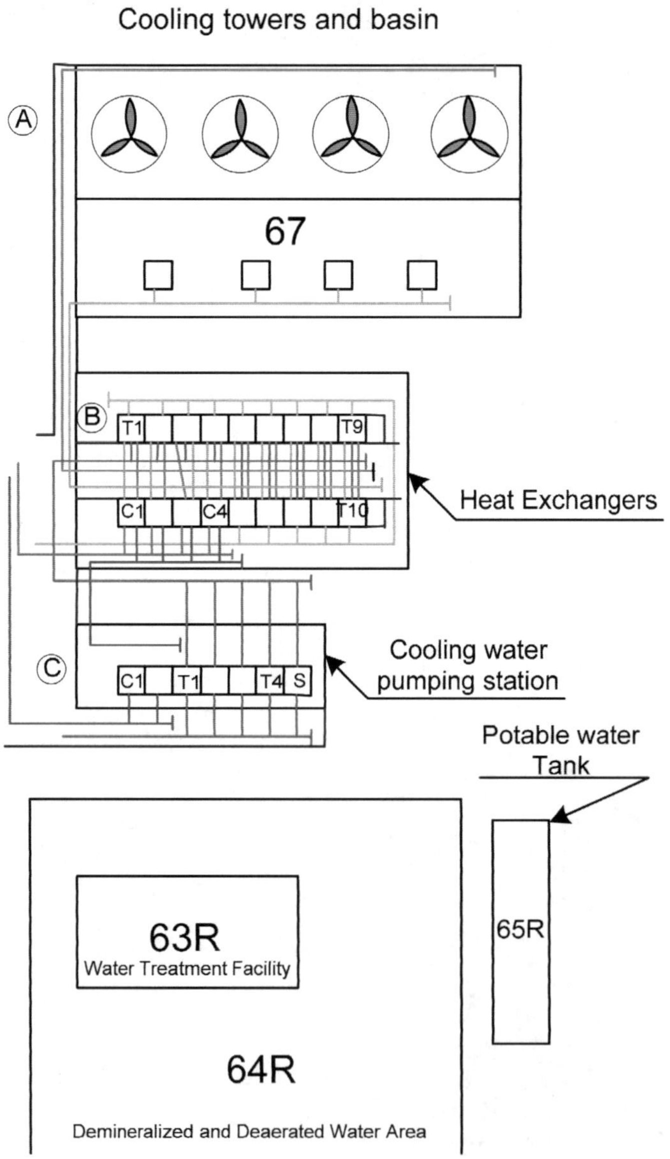

Fig. 4 – Layout of the HRS

Conclusions

The ITER Cooling Water System has been recently modified by introducing two closed intermediate Component Cooling Water Systems (CCWSs) to transfer the heat generated in the Plasma and in its auxiliary equipments to an open Heat Rejection System (HRS).

These design modifications will give as main benefit the minimization of the capital and operative costs, the optimization of the piping layout and the reduction of its length, the standardization of the CCWS pump and heat exchanger units.

Another beneficial effect is because the closed CCWS represents a second containment barrier against the possible release of contaminants, both in gaseous and liquid form, from the primary heat transfer nuclear systems to the environment.

The environmental impact of the Cooling Water System of the ITER Reactor

G. Dell'Orco, W. Curd, D. Gupta, L. Fan, K.P. Chang, M. Glugla, B. C. Na, M. Benchikhoune

International Thermonuclear Experimental Reactor Organization (ITER-IO)

13108 Saint Paul lez Durance, France

Tel:+33-(0)4-42-25-77-60,Fax:+33-(0)4-42-25-26-00

Abstract

ITER is the largest fusion experimental reactor in the world designed to reach the first plasma at mid of 2018. ITER is designed to demonstrate the technological feasibility of the nuclear fusion energy conversion, at plant scale, from high temperature Deuterium-Tritium plasma using the TOKAMAK magnetic confinement arrangement and with a power amplification at least of 10. The ITER design experience, the effort in developing the leading-edge technologies and the knowhow acquired during its operation will guide the realization a proper long-term R&D fusion programme. ITER will bridge the Nuclear Fusion toward the large scale commercial production of electricity at competitive cost with other sources by 2045.

ITER is provided with a Cooling Water System (CWS) designed to rejects to the environment all the heats generated from the components (nuclear and non nuclear) using water as coolant in Primary Heat Transfer Systems (PHTSs). The only exception is the Vacuum Vessel (VV) whose heat is released, via a separate heat transfer system, to the air coolers. The total heat to be released to the environment during the DT pulse is about 1270 MW by water via the Heat Rejection System (HRS) and Cooling Tower System (CTS) plus about 13 MW mainly from the VV (10 MW) and other systems (3 MW) by air coolers.

Among others, one of the main critical issues of the ITER reactor is the minimization of the releases of all the gaseous, liquid and solid effluents from the PHTSs to the environment.

This paper describes these main releases from the CWS (PHTSs, HRS and CTS), both during normal operations and main periodical maintenance.

ITER Cooling Water System

All the heat generated in the Plasma, during the D-T reaction, will be transferred, through the Tokamak Cooling Water System (TCWS and PHTSs), to the intermediated closed loop Component Cooling Water System (CCWS) and then, via the open loop Heat Rejection System (HRS) to the environment.
The HRS also absorbs heats through the CCWS from other non nuclear systems like the Chilled Water System (CHWS), the Cryogenic System, the Steady State Electrical Power Network (SSEPN) and other auxiliary systems. The HRS rejects to the environment all the heats from the ITER components (nuclear and non nuclear) with the only exception the Vacuum Vessel whose heat are released via a separate primary heat transfer system to the air coolers.
In Fig.1 is shown the main organization of the ITER Cooling Water Systems

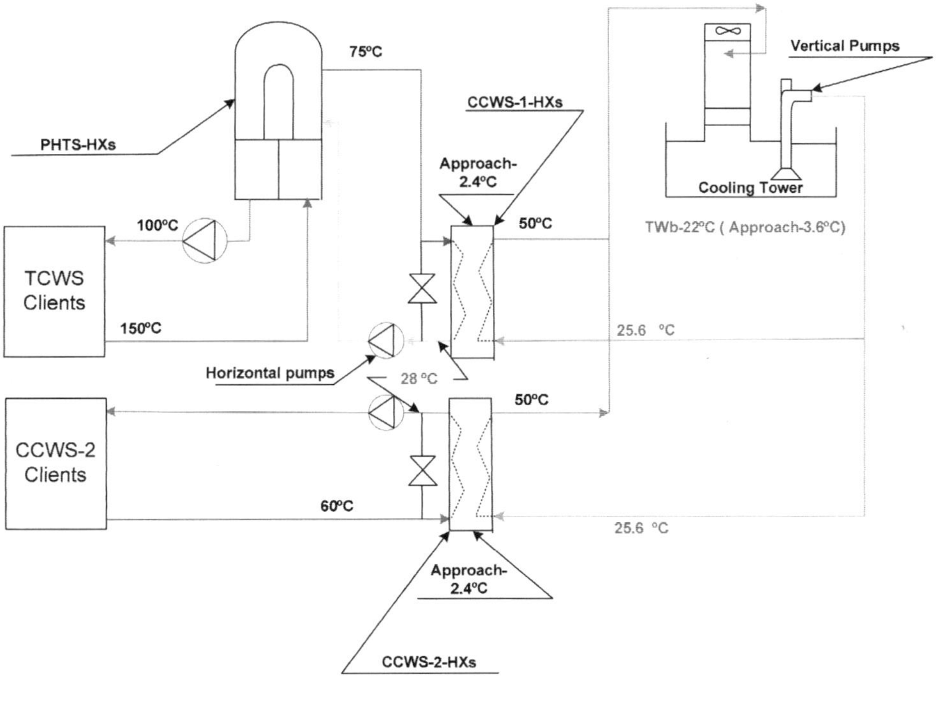

Fig. 1 – ITER Cooling Water System

The Tokamak Cooling Water System (TCWS)

The TCWS consists of PHTSs, the chemical and volume control systems (CVCSs), the draining and refilling systems, and the drying system.

During plasma operation, the PHTSs transfer heat generated in the components to environment via the intermediate closed component loops Cooling Water Systems (CCWSs) to the Heat Rejection System (HRS) by the Cooling Tower Systems (CTS).

In between pulses the PHTSs remove decay heat from the plasma facing components (First Wall Blanket and Divertor) as well as from the Vacuum Vessel.

There are 7 PHTS loops as follows:

- 3 loops for the First Wall and Blanket (FW/BLK) PHTS,
- 1 loop for the Divertor and Limiter (DIV/LIM) PHTS,
- 2 loops for the Vacuum Vessel (VV) PHTS, and
- 1 loop for the Neutral Beam Injector (NBI) PHTS.

The VV PHTS has a safety role in that it provides the ultimate decay heat removal even when the other PHTSs are not available during some very severe accidental events.

The safety role of the other PHTS loops for in-vessel components is limited to the confinement of radioactive inventory. The containment walls of the in-vessel PHTS loops are safety important class (SIC) because of their confinement function, however the active function of the PHTSs is non-SIC except some insulation valves.

The introduction of this further CCWS/1 closed loop permit the control of the possible releases of gaseous and solid nuclear products from the TCWS to the HRS. Therefore, the secondary water loop is daily controlled to prevent any normal and abnormal release of contaminants to the HRS.

The main PHTSs are provided with the Chemical and Volume Control System (CVCS), one for each in-vessel component PHTS (FW/BLK, DIV/LIM, and NBI), whose function is to control the water chemistry of the PHTSs, to manage the level control in the PHTS loops, to provide the main pump cooling, and to provide spray flows for pressurizer pressure control.

Each CVCS unit contains inlet/outlet mechanical filters to trap the Activated Corrosion Products (ACP) produced in the water loops by disposable stainless steel cartridges.

The CVCSs are also provided with Ion Exchange Resins bed demineralisers. Both the mechanical filters and the resin bed demineralisers are surrounded by a radiation shield, due to the high contamination of the trapped/absorbed ACPs.

The Heat Rejection System (HRS)

The HRS provides the final heat sink and rejects all heat loads in the ITER plant with the exception of the Vacuum Vessel.
The main HRS clients are, Fig 1: i) CCWS/1 or the secondary circuits of the main Primary heat Transfer Systems (PHTS) of the TCWS, ii) CCWS/2 or the secondary circuits of the CHWS, the Cryogenic System, the Steady State Electrical Power Network (SSEPN) and other auxiliary systems.
The HRS consists of the Cooling Tower System (CTS), with 4 air forced draft main units, the water basins, the Vertical Pumps together with the associated water make up and blowdown systems. Whilst the CTS transfers the heat from the CCWS to the atmosphere, the water basin and its ancillary circuits have the following objectives:

1) Provides buffer heat capacity by proper basic water storage to average the heat load during pulsed operation.
2) Provides makeup and blowdown flows to perform the chemical control of the salt concentration in the water basin and to prevent the growth of biologic species (i.e. algae).

Releases from the CWS

Solid waste release
The main solid radwaste release from the CWS (namely from the CVCSs) are the mechanical filters and the spent ion exchange resins from the demineralisers with high contamination of the trapped/absorbed ACPs.
All these units will be routinely changed every year with an overall amount of about 12 m3/y with a T activity < 2 TBq/m3 and ACP activity < 0.3 GBq/m3.

Liquid waste release

After 10 y of operation in Deuterium/Tritium phase the T concentration in the 800 m3 of water of the FW/BLK and DIV/LI PHTSs will reach the value lower than 1.85 TBq/m3. The ACP contains will be negligible because trapped in the CVCS loops.

This water will be stored on site to the properly reprocessed before the final release to the environment.

The water in the secondary CCWS/1 will be periodically released according to the T limit concentration imposed to the French Local Authority (Prefecture).

Liquid and Gas release from the CTS

When the CTS will operate at full power (450 MW) the water evaporation rate will be about 200 kg/s. The water make up unit will correspondently provide 260 kg/s from the river Verdun because about 60 kg/s will be released as blow down to control the salt concentration in the loop and to prevent the growth of biologic species (i.e. algae).

The liquid effluents will be release in the river La Durance, after having performed the routine control of the main parameters, shown in Table I, in the storage basins (four each of 3000 m3) as authorized by the French Local Prefecture.

Parameter	Authorized limit	Max daily authorized limit	Periodicity of control
Total Flow rate effluent (m3/d)	4000 daily peak	3000 monthly average	continuous
PH	5.5 - 9		continuous
Max. outlet temperature (C)	30		continuous
Chlorates (mg/l)	200	450 (kg/d)	daily
Sulfates (mg/l)	700	1575 (kg/d)	daily
Phosphates (mg/l)	10	22.5 (kg/d)	daily
Fluorites (mg/l)	1	2.25 (kg/d)	daily
Tritium (MBq/m3)	10	22.5 (GBq/d)	daily
Total radionuclides (MBq/m3)	0.1	225 (MBq/d)	daily

Table I – Limit of authorized water release and chemical parameters

Conclusion

The ITER Cooling Water System is based on some Primary Heat transfer Systems (PHTS) and two closed intermediate Component Cooling Water Systems (CCWSs) that transfer the heat generated in the Plasma and in its auxiliary equipments to an open Heat Rejection System (HRS).

This organization permits a full control of the water coolant, its relevant contamination due to the neutron generated in the Plasma by DT reaction and the related effluent management.

In fact, the closed CCWSs represents second barriers against the possible release of contaminants, both in gaseous and liquid form, from the primary heat transfer nuclear systems to the environment.

The adoption of the Cooling Tower also permits the reduction of the water release in the river close to the site.

150d Processing Requirements for Iter Tokamak Exhaust Processing

James E. Coons, Los Alamos National Laboratory, P.O. Box 1663, Los Alamos, NM 87545, R. Scott Willms, Group Leader, Los Alamos National Laboratory, P.O. Box 1663, Los Alamos, NM 87545, and Bryan J. Carlson, Project Manager, Los Alamos National Laboratory, P.O. Box 1663, Los Alamos, NM 87545.

The system and interface requirements for the ITER Tokamak Exhaust Processing are complex and stringent. Investigation into the input streams and output constraints yielded interesting information. A summary of the results of the investigation will be presented.

228a Developing Fusion Power: Using Plm Collaboration to Manage Engineering Content

Eric Martin, Desgin Office, ITER IO, Cadarache Joint Work Site -Building 519, 13108 St Paul-Lez-Durance, Cadarache, France

After a brief overview of the ITER Project, the presentation will cover the ITER Design Office challenges, including in particular a large number of inter-connected disciplines in mechanical and plant design, a complex interface system as well as a demanding collaboration scheme to allow the ITER Partners to remotely contribute to the Project an integrated manner.

Then the presentation will focus towards the solutions that are already in place, including key processes and procedures, the CAD and associated integrated data-base in the area of mechanical and plant design as well as configuration control. The collaboration and the simulation of the machine assembly and maintenance operations will be only introduced as handled by separate presentations at this conference.

Then the presentation will focus towards on-going and up-coming developments in particular in the area of plant design and engineering data-base allowing the connection of CAD data to non CAD data.

228b ITER Design Collaboration

Eric Thomas, Design Office, ITER IO, Cadarache, France

Starting from the Context of Design Activities already depicted by the previous session (E. Martin, ITER DO Head), this presentation will report on the decisions and implementation of Design Office processes and aimed at providing a project wide environment which facilitates distant collaboration:

This will include in Particular:

- A review of the driving factors leading to the Choice of a Product Lifecycle Management System, with emphasis on collaborative aspects.

- The construction of 3 schemes of collaboration covering the wide range of industrial cases found in ITER

- The Implementation of technical and Quality enablers supporting the ITER DO collaborative approach

The Presentation will conclude with some general observations on the experiences and forthcoming challenges in developing a Collaborative Design Office structure.

228c ITER Assembly Simulation

Stephen Murray, Design Office, ITER IO, Cadarache, France

Design and manufacture of ITER's reactor and ancillary equipment is performed by widely dispersed, collaborating teams. The IO central team will be responsible for on-site assembly, installation and maintenance.

To guarantee the assembly and maintenance schedules will require a high degree of confidence in the associated procedures and hardware, which must be developed in parallel with component design. Since a physical mock-up is largely ruled out by the the scale of the equipment, these processes will be developed and verified using CAD simulation.

ITER has deployed simulation tools which are integrated with the design office CAD system and PDM, to leverage the existing digital mockup. Kinematic equipment, such as tools and robots, are built on existing 3D data and assigned 'tasks' in process simulations. Human activities can also be modeled. Simulations are based on process plans, either developed interactively during the course of a study or imported from an external source, for validation of pre-defined processes. Detailed analysis and optimization of procedures are performed using tools for clash/clearance detection and path-planning.

228d Design Collaboration Methods for the U.S. ITER Project

David E. Williamson, Oak Ridge National Laboratory, PO Box 2008, Oak Ridge, TN 37831-6483

The United States is part of an international collaboration to construct and operate ITER, an experimental device that will demonstrate the scientific and technological feasibility of fusion energy. ITER will be constructed at Cadarache, in southeastern France. One of the more challenging aspects of the project is the need to perform concurrent engineering of multiple complex mechanical systems that are to be manufactured and installed by different Partners. In order to provide a common framework for all phases of design and construction, the Project is developing a distributed Product Life Cycle Management (PLM) system for CAD data, with the capability to manage multi-group work flow and data approval within several integrated project teams. In addition, we are investigating the use of advanced collaboration technology, such as 3D immersive displays, to improve the quality of design integration activities.

ITER Plant Design Challenges

Seokho H. Kim
US-ITER, Oak Ridge National Laboratory
1055 Commerce Park
Oak Ridge, Tennessee 37830-6483
kims@ornl.gov, (865) 241-1392

US ITER participates in International Thermonuclear Experiment Reactor (ITER) that is constructed in Europe, at Cadarache, France. Specifically, US-ITER is responsible for design, engineering and procurement of the Tokamak Cooling Water System (TCWS). Because ITER as an experimental facility is designed and built first time, various design ideas and new concepts are evolving every day. Therefore, it is challenging that the design and engineering of TCWS is conducted in a manner to accommodate rather quickly such frequent changes in client system design while not to delay the overall project progress. Another difficulty exists in identifying physical and functional interfaces of TCWS with other systems. Details in system design and engineering challenges will be discussed during the presentation.

294a Water Chemistry and Corrosion of ITER Tokamak Cooling Water Systems

Otakar Jonas and Lee M. Machemer. Jonas, Inc., 1113 Faun Rd., Wilmington, DE 19803

This paper briefly discusses the ITER Tokamak Cooling Water System design and operation and the technical approach used in the selection and evaluation of the piping materials. Dual certified 316/316L low carbon austenitic stainless steel is recommended for the cooling water system piping. The remaining challenges include presence of copper and CuCrZr alloy at water flow velocity up to 12 m/s, uncertainty in the evaluation of the water chemistry and corrosion potential during neutron flux (O2, H2O2), crevice corrosion of 1000 tons of borated and ferritic steel plates, and the effect of high magnetic fields on the corrosion and deposition processes and the forces generated by interaction with delta ferrite.

294b Modeling of Activated Corrosion Product Transport: A Theoretical Approach

Stanislav I. Golubov[1], Jeremy T. Busby[2], Roger E. Stoller[1], and Andrei Petrov[1]. (1) ORNL, P.O. Box 2008 MS6138, Oak Ridge, TN 37831-6138, (2) Oak Ridge National Laboratory, PO BOX 2008 MS6138, Oak Ridge, TN 37831-6138

Within the ITER machine, activated corrosion products (ACPs) will be present in the various in-vessel and vacuum vessel coolant loops as well as in any coolant loops related to test modules, auxiliary heating or diagnostics equipment. The activation products will impact occupational exposure, routine effluents to the environment, and potential releases during accidents. This fact has made the ACP inventory evaluation an important task for ITER public and occupational safety. Thus modeling of ACPs for ITER operation is a necessary requirement providing control and prediction of ACPs during operation. Such modeling has been done so far using the so-called PACTITER code, which is a modification of the PACTOLE code used for light water reactors and is based on a finite element approach where the primary circuit is represented by an arrangement of several volumes in which transient mass balance equations are solved. In the framework of the model there are a finite number of such volumes which are connected to each other by additional mass exchange equations. Therefore, the code provides an approximate solution with the accuracy depending on the number of the control volumes. However the model is limited to solution by numerical methods which may restrict the accuracy of such calculations under complex ITER operation scenarios. In this work we show that the problem can be described in the framework of a continuity equation for laminar flow of water in the coolant circuit. The continuity equation is written in a form which accounts water filtering and precipitation of the ACPs along the cooling ring and can be solved by both analytical and numerical methods. Analytical solutions for the case of a simple ring describing ACPs accumulation in the ring and filters are presented and discussed.

Studsvik

A Review of Corrosion and Water Chemistry Aspects Concerning the Tokamak Cooling Water Systems of ITER

Anders Molander
Studsvik Nuclear AB
SE-611 82 Nyköping
Sweden
anders.molander@studsvik.se

Introduction and outline– Review of corrosion and water chemistry of ITER performed

- Summary of the review of some aspects concerning
  - the materials behavior
  - water chemistry and
  - water side corrosion in ITER environment.
- Goals of the review
  - Identify critical corrosion aspects for ITER.
  - Summarize the system design from a corrosion engineering perspective.
  - Point our relevant corrosion mechanisms important for the systems.
  - Justify requirements concerning corrosion and water chemistry.
  - Point out any lacking knowledge and any need for additional work.
- Summary of key issues identified which need further considerations and investigations to assure the operational lifetime of ITER.

Studsvik

Proactive work – Experience from LWRs

Lessons learned from LWRs

- Water quality
- Sensitized Type 304
- IASCC
- Crevices
- Condensers
- SCC of Type 316L

Corrosion-Related Capacity Factor Losses in BWRs

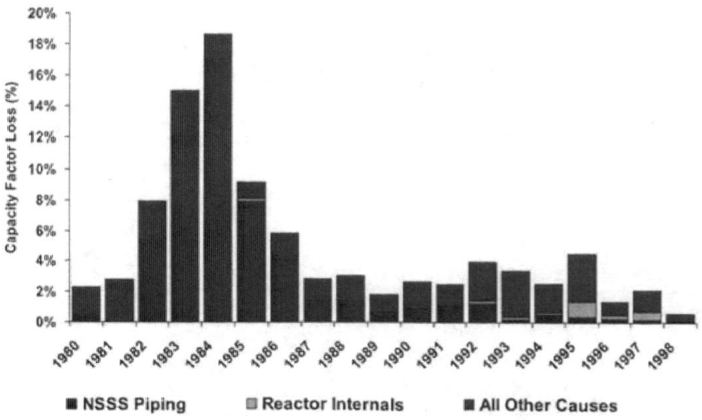

12

EPRI

Studsvik

The Tokamak Cooling Water Systems of ITER

First wall and blanket

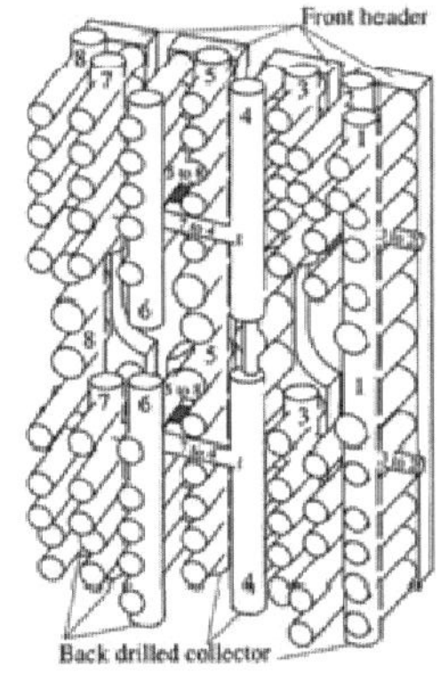

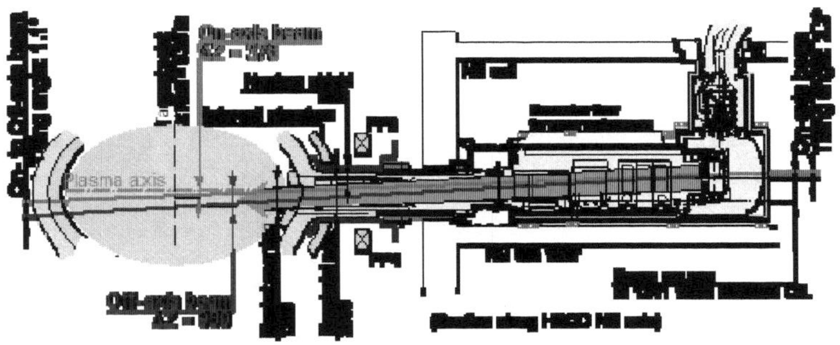

Neutral beam injector

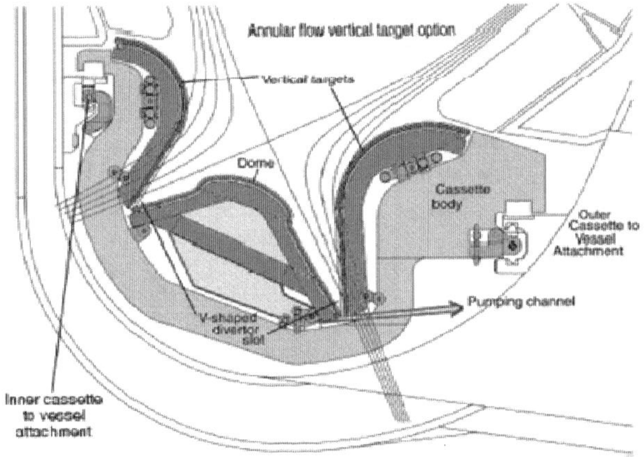

Diverter and limiter

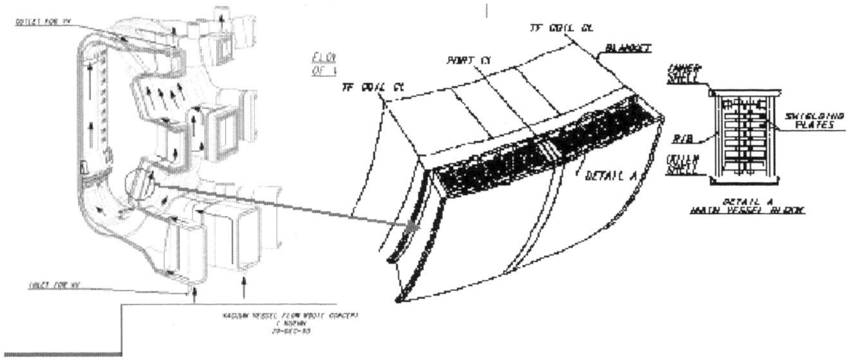

Vaccum vessel

Studsvik

Environment

- Pure water at temp lower than BWRs
- Temperature close to ambient →100-150°C →240°C
- Irradiation: Radiolysis and hydrogen addition.
  - Hydrogen is planned to be added to ITER systems (except VV) to achieve reducing conditions
  - But ITER conditions less favorable for radiolysis suppression than for LWRs.

Studsvik

Radiolysis aspects

- Hydrogen decreases the radiolytic oxidant concentrations significantly but only down to a plateau level

  The residual oxidant amount is enough to create high corrosion potentials

  Radiolysis suppression as in LWRs is not possible to obtain in ITER PHTS according to **EU studies**.

  → Oxidizing conditions will prevail

  → Corrosion potential will be high

  Hydrogen is still recommended to decrease the oxidant formation.

- According to a **Russian study** reducing conditions are possible to obtain.

Studsvik

Water chemistry specification

- The ITER specification was set in the 90ies from BWR guidelines at that time

Required Operational Parameter[2]	Feedwater	Upper limits for action
Conductivity (at 20°C), μS/cm	<0.1	<0.3
Oxygen, μg/kg	<100	<10[1]
Chloride and/or Fluoride, μg/kg	<0.5	<5
Sulphate, μg/kg	<20	<5
Copper, μg/kg	<0.5	<5
Iron, μg/kg	<1	<5
Hardness (Ca, Mg, etc), μg/kg	<5	<5
Oil products, organic, μg/kg	<100	<100
pH at RT	neutral	-
Hydrogen added to suppress radiolysis (cm3/kg @ STP)	25	-

[1] The oxygen content for the vacuum vessel cooling system will be limited <100 μg/kg.
[2] A plasma pulse will not be initiated if any of the above limits is foreseen to be exceeded during the course of the pulse itself.

BWR Mean Reactor Water Conductivity History (Provided by GE Nuclear Energy)

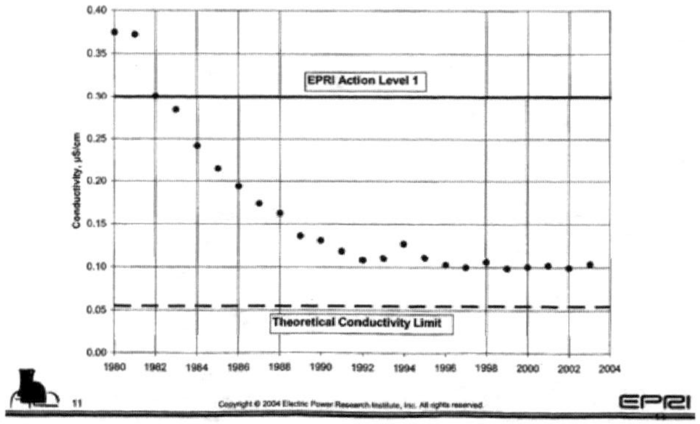

- Revised target values and action levels were tentatively suggested for each PHTS.
- Situation for VV: Hydrogen addition and CVCS system should be considered.

Studsvik

Importance of keeping high purity water

- Low levels of impurities affect several types of corrosion
  - Stress corrosion cracking of stainless steels
  - Crevice corrosion of stainless steels
- Low levels of impurities affect can accumulate in narrow crevices causing
  - Aggressive local solutions in crevices/low flow regions
  - Increased local corrosion rates
  - Corrosion during drying
  - Increased stresses through the formation of corrosion products
  - Stress corrosion cracking

Water purity

Tentative system water target values

Parameter/PHTS	PFW/BLK	DIV/LIM	NBI	VV
Conductivity, µS/cm	<0.1	<0.1	<0.1	<0.1
Sulfate, ppb	<5	<5	<5	<5
Chloride, ppb	<5	<5	<5	<5
Copper, ppb	<0.5	<2	<0.5	<0.5
Hydrogen, Ncm^3/kg	25	25	25	- -?
Oxygen, ppb	-	-	-	<100?

Based on present BWR mean values and guidelines.

Studsvik

Additional recommendations

- Specifications should clearly refer to the system water.
- Defining CVCS outlet water (feedwater) or make-up water for VV not enough.
- Define monitoring points and methods.

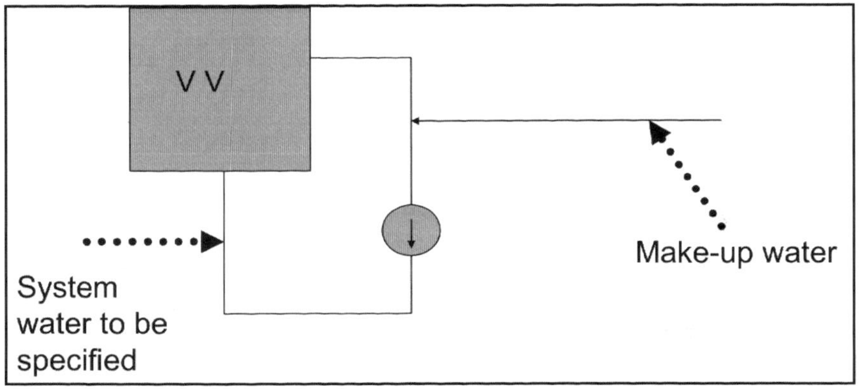

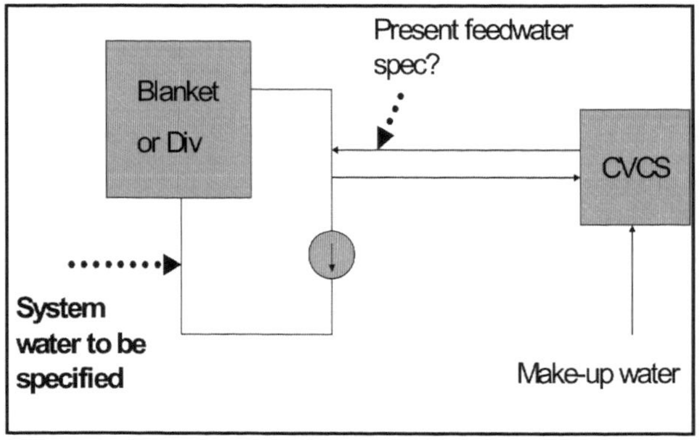

Studsvik

Summary of water chemistry concerns

To resolve the radiolysis situation is a key issue for predicting ITER corrosion conditions.

To design for high water purity is essential to avoid corrosion attacks.

Hydrogen addition and CVCS system for the VV should be considered.

Define monitoring points and methods.

Corrosion properties of ITER materials

Type 304

- IGSCC of sensitized stainless steel
- The most common and most familiar corrosion phenomena in BWRs
- Caused early significant power losses and repair costs
- Now well understood and mitigation methods are available

Corrosion properties of ITER materials

Type 316L(N)-IG

- Through the introduction of L-grade materials in BWRs it was believed that the probability for future cracking incidents would be very low. However IGSCC reported from several countries
- In most cases cracking seems promoted by a cold deformed/hardened surface layer caused by grinding or other type of machining. Mechanical factors important like the residual strain close to the weld fusion line and high local tensile stresses at the bottom end of the weld which usually is in contact with reactor water
- For ITER cold working and post weld treatments are considered as an important remaining issue.
- Non-sensitized stainless steel in annealed conditions should not show SCC in pure reasonably oxidizing water as in ITER

Corrosion properties of ITER materials

XM-19 stainless steel (22Cr-13Ni-5Mn-2Mo-0.3N)

- XM-19 XM-19 was developed as a high strength corrosion resistant alloy
- Good experience in BWRs
- Stress corrosion properties have been found excellent for XM-19, but tests also show that SCC can occur, especially for high stresses in oxygenated conditions. Crevices enhance the attacks as for Type 316L materials.
- In summary, the corrosion properties in general should be better for XM-19 than for Type 316L and the dependence on different parameters such as temperature, crevice conditions and irradiation should resemble those of Type 316L.

Corrosion properties of ITER materials

Borated stainless steels

- Most literature indicate slightly worse corrosion behavior than for Type 304 in non sensitized condition and slightly better behavior in sensitized condition.
- Pitting and crevice corrosion resistance seems slightly worse than for Type 304.
- The general corrosion rate should be similar or marginally higher than for Type 304.
- Irradiation effects have not been covered as the neutron flux is low inside the vacuum vessel.
- Crevice corrosion concerns, see below.

Corrosion properties of ITER materials

Ferritic steel Type 430

- Literature search gave only limited relevant data.
- Type 430 should withstand the nominal environment of the VV.
- Type 430 is considered to have the lowest corrosion resistance of the ITER steels but still good enough for pure water at 100°C if not sensitized.
- From experiments as a TSP material for PWRs Type 430 showed as good behavior as Type 347 in the bulk environment. In crevice environment the behavior was worse.
- Crevice corrosion concern, see below

Studsvik

Corrosion properties of ITER materials

Copper alloys

- Reducing conditions are said to be needed to protect Cu from corrosion
- Radiolysis work indicates that reducing conditions cannot be established in irradiated system parts
- High corrosion rates >10 μm/year will be obtained according to tests during such conditions
- Low risk for galvanic corrosion in pure oxidant-free solution but possible in oxidizing environment
- From a corrosion point of view Cu behavior seems as a serious concern

Studsvik

Some important corrosion issues with focus on stainless steels

- Effect of temperature
- Crevice corrosion
- Effect of impurities
- Effect of Cu on SS corrosion
- IASCC

Studsvik

Temperature dependence - BWR data

Initiation slower at lower temperatures

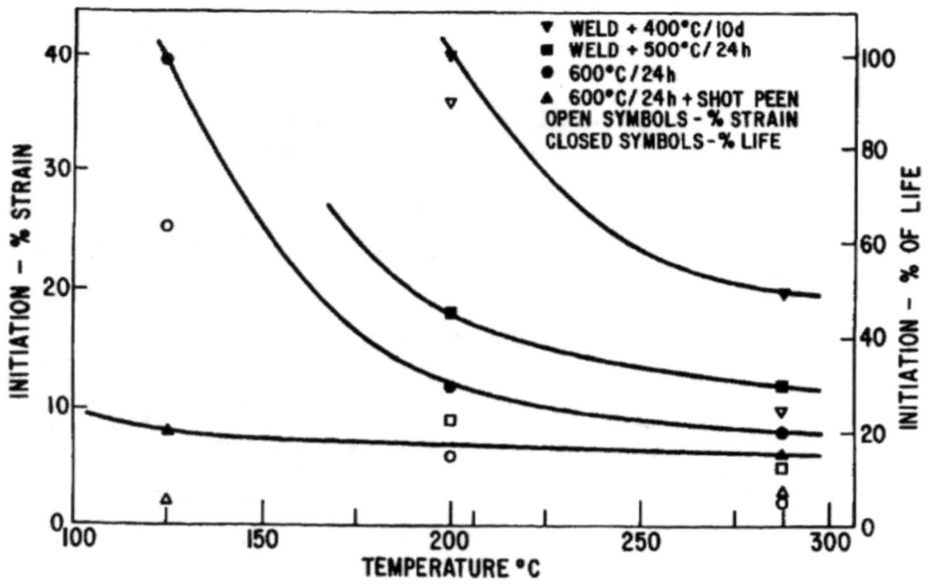

High CGR in un-pure water at about ITER temperature

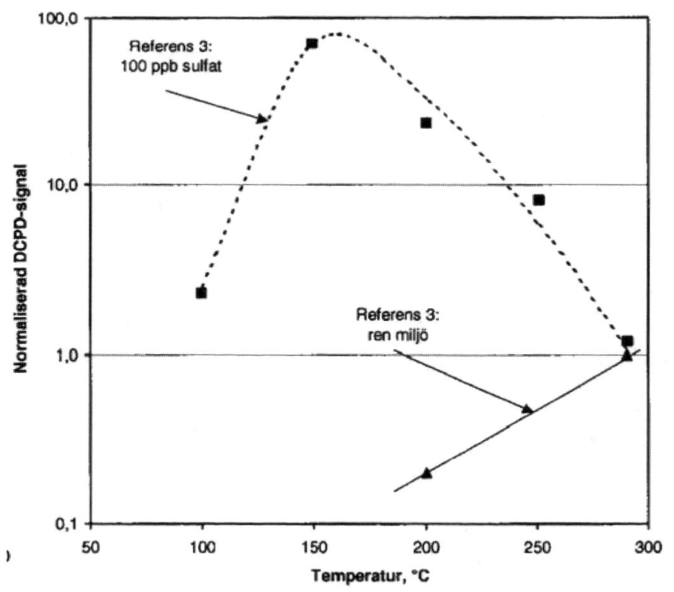

Studsvik

Effect of temperature on SCC

- Few data and some contradictory data makes it difficult to use the data for ITER predictions
- But initiation retarded with decreasing temperature
- Some propagation data shows a max at about ITER-temp, other data show a monotonic decrease with temp.
- Initiation is promoted and propagation increases with conductivity (poor water chemistry), increasing ECP and presence of cold work.

Advice to BWR plants

Poor water chemistry at any temperature will cause significant SCC. A plant should do everything possible to identify and fix problems before they occur.

Should also apply for ITER…

Studsvik

Crevice corrosion and crevice enhanced IGSCC

- Crevices are known to accelerate crack initiation both in sensitized and non-sensitized material
- In plant failures in BWRs have occurred
- In crevices different species can be concentrated with factors of hundreds or thousands
- In the crevice dissolution of metals to positive ions will draw negative ions into the crevice
- Thus, a more aggressive usually acidic crevice environment will form and keep the crevice attack going

Conditions in the VV

- As the flow conditions in the VV are rather slow, corrosion products and impurities can accumulate in creviced close to stagnant areas.
- During such conditions more a aggressive environment can be formed as the metal ions attract negative ions like chlorides.
- If so Type 430 is susceptible to local corrosion like crevice corrosion and pitting.

Shielding blocks in VV

High number of crevices in low flow environment

Limited corrosion data base for the given materials

Testing suggested

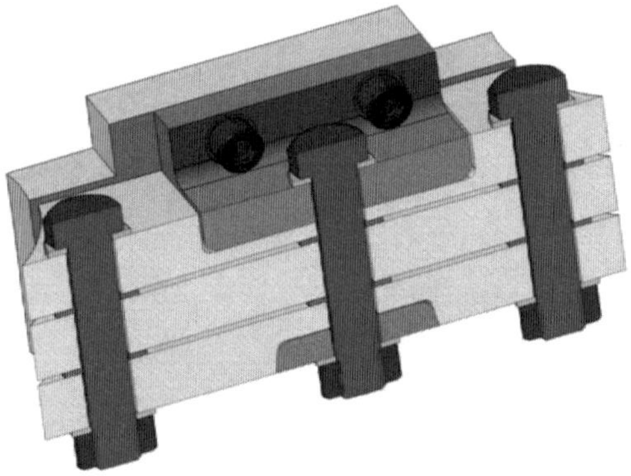

Studsvik

Crevice attacks promoted by drying/wetting

- During drying impurities are trapped in low flow areas and crevices
- More aggressive solutions formed
- During refilling deposits dissolve and again an aggressive solution can be formed
- Repeated drying/wetting can cause corrosion, in particular crevice corrosion
- Tests performed show attacks even on Type 316L(N)-IG
- Difficult to define a realistic concentration factor but from literature factors of hundreds and thousands are well-known

Studsvik

Corrosion experiments during simulated drying conditions

- Shutko, Belous and Akimkin found severe corrosion during testing in 1M NaCl + 1 M Na_2SO_4 at 120°C already after 65 hours.
- Lorenzetto, Helie and Molander found also severe cracking at temp >300°C. The concentration factors were 100 to 1000.
- Tähtinen also found SCC and severe corrosion on CuCrZr in an aggressive solution.
- In all cases the environments might have been too corrosive but the results indicate the importance of the drying procedure not only for shielding plates.

Studsvik

Crevice Corrosion - Summary

- Crevice corrosion in shielding block material is considered to be a potential risk for the VV-system
- Corrosion tests for VV are suggested:
  - to establish an acceptable concentration factor
  - to estimate possible concentration factors that can be obtained
- Other locations in all other systems should be identified and considered
- A realistic, not over conservative, test method is needed.
- Conditions during drying should be taken into account.

Studsvik

Influence of impurities on IGSCC

- Ingress of sulfate and chloride are well know to increase the SCC CGR rate of Type 304 and to decrease the initiation time
- Transients and increased conductivity can have a tremendous impact on the estimated total crack growth rate during a reactor year
- Should in general be true also for Type 316L(N)-IG
- At 5 ppb no major detrimental effect has been detected from either sulfate or chloride in lab tests
- At 30 ppb an obvious acceleration is obtained
- After an impurity ingress an increased CGR can remain for an extended period
- Water chemistry extremely important during all phases of operation

Studsvik

Effect of Cu on SS corrosion

- Presence of Cu is known to have certain detrimental effects in LWRs
  - BWR fuel cladding
  - PWR steam generator corrosion
- Rather high Cu contents needed to influence SCC behavior in Type 304.
- Some BWRs are running with copper contents up to ppb level without any evident effects compared to non-copper BWRs.
- Same concentrations of Cu as in BWRs (copper plants) are specified for ITER.
- If such low concentrations are possible to maintain in the divertor system is however uncertain
- Running the divertor system with high oxidant contents will produce unfavorable conditions for SS with a risk for localized corrosion as a consequence.

Studsvik

IASCC concerns

- The maximum dose of ITER components
  - is above the threshold for IASCC in oxidizing LWR environments
  - is in the same range as the threshold for IASCC in reducing LWR environments
- The time after obtaining the critical dose is limited
- ITER temperature is lower than for LWR which should delay initiation but the influence of temperature on IASCC as well as SCC is unclear.
- The time and the temperature are regarded as beneficial factors but knowledge of the redox conditions and the ECP considered essential to better estimate the IASCC risk.

Summary of concerns

- Radiolysis
- Cu- alloys behavior in all aspects in slightly oxidizing environments
- Crevice corrosion
  - Shielding blocks
  - All other system during **drying**/wetting conditions
- Welds and cold work in Type 316L(N)-IG
- IASCC – Redox conditions?
- Water purity – Monitoring – Cu-contents

Studsvik

Studsvik

Role of Water Treatment in Corrosion Control of ITER Cooling Water System

Frances Cutler, Consultant

Andrei Petrov, Engineer, US-ITER

ABSTRACT

ITER is an experimental Tokamak fusion reactor that has been developed in collaboration with seven international partners, including the United States. The ITER cooling water system (CWS) is an integral part of the Tokamak reactor complex and is designed to perform the following key functions:

* Remove heat from in-vessel components during the fusion process
* Control the fluid thermodynamic parameters
* Maintain water chemistry within acceptable limits

To perform these functions safely and reliably, the CWS requires an adequate supply of high-quality demineralized and deaerated water since corrosion mechanisms are frequently related to water purity. It should also be noted that the chemistry environment is an important factor in certain fatigue failure mechanisms. Unsatisfactory water chemistry can also lead to deposition of solids that could permit underdeposit corrosion and increase the level of activated corrosion products formed. This paper discusses the influence of makeup water purity on cooling water chemistry and the criteria for makeup water specifications. Design features of the initial water treatment facility, alternative methods for supplying makeup water, and problems and issues of concern relative to the makeup water treatment system are summarized.

BACKGROUND

Tokamak operation and maintenance requires draining and refilling of the cooling water system as well as additional makeup water to compensate for losses due to leakage. Water quality requirements for the Tokamak cooling water system have been studied by experts in a number of fields such as metallurgy, radiation, corrosion, and water treatment. The resulting consensus is that strict control of water quality with respect to oxygen and dissolved solids such as chloride and sulfate is required to minimize corrosion and deposition. Consequently, the initial cooling water specifications were re-examined and were subsequently made more stringent resulting in re-evaluation of the design for the makeup water storage and treatment systems.

RELATIONSHIP BETWEEN MAKEUP WATER QUALITY AND THE GENERATION AND DEPOSITION OF CORROSION PRODUCTS

A number of failure mechanisms and other availability problems have been identified as being directly influenced by the generation, transport, and deposition of corrosion products. In a radioactive environment, it is even more critical to minimize the generation of activated corrosion products. Tight control over the purity of the cooling water and selection of the appropriate metallurgy are essential to maintaining the appropriate environment to minimize corrosion.

Following proper pre-operational cleaning and start up of the system, causes of chemistry excursions in the cooling water of the Tokamak would primarily result from contamination of the makeup water or the storage tanks within the system. The actual types of contamination associated with makeup treatment systems will vary depending on the source water composition and makeup system design, but typically include sodium, chloride, sulfate, silica, oxygen, carbon dioxide and organic carbon. Chloride and oxygen are the constituents most frequently associated with inducing corrosion while sulfate and silica are associated with deposition.

The magnitude of the influence of makeup water quality on cooling water chemistry depends on a number of factors:

* Makeup rate
* System water quality requirements
* Anticipated number of start-ups
* Presence or absence of in-line polishers in the cooling water system

Criteria for Makeup Water Specifications

Makeup Rate In systems where the rate of makeup is less than 1% and concentration within the system is not a problem, the quality of the makeup water may be less stringent than the water chemistry requirements for the operating system. For example, if the chloride limit for a system with a makeup rate of <1% is 0.1 ng/kg (ppb), the chloride limit for the makeup treatment plant may be between 1-10 ppb although an upper limit close to 10 ppb would provide no safety margin.

Cooling water systems are rarely free of areas where concentration of contaminants can occur. For this reason, even a system with less than 1% makeup should consider limits no more than 50% of the maximum allowable to meet in service requirements. The higher the rate of makeup to the system, the closer the limit on makeup water quality should be to in-service water quality requirements.

System Water Quality Requirements In-service system water quality requirements determine the quality of the makeup water that should be sent to the system. In general, in-line polishers within the cooling water system should not be relied upon to treat the makeup water, but should be used as "insurance" in the event of an excursion in quality from the makeup water treatment system.

Anticipated Number of Startups For system startup, the makeup water quality should meet in-service water quality requirements unless there are in-line polishers that can be used to treat the makeup water. However, even in this case, the quality of the makeup water should be sufficiently close to the required quality so that the polishers are not exhausted prematurely. The greater the number of anticipated startups, the more important it is to have the quality of treated makeup water meet in-service requirements.

Experimental systems such as the Tokamak are likely to have frequent startups, some of which may follow extended outages. In this type of circumstance, sufficient demineralized water storage for preliminary fill up and drains would reduce the demand for high purity water required for the final fill.

Presence of In-Line Polishers The presence of full in-line condensate polishers should not be used as justification for relaxing makeup water quality requirements. Since the current design of polishers for the Tokamak cooling water system is for ~1% polishing, it is even more important <u>not</u> to relax makeup water quality requirements or rely upon in-service polishing.

MAKEUP WATER TREATMENT SYSTEM

Design Features of the Initial Water Treatment Facility

The design features of the initial water treatment system were for an ion exchange demineralizer with decarbonator. Interim modifications included reductions in makeup demineralizer flow and greater reliance on tanker delivered water. It is believed that there are a number of preferable demineralizer designs that would better meet the needs of the Tokamak cooling water system.

Alternative Methods for Supplying Makeup Water

A number of alternative makeup demineralizer designs are available that would provide both the quality and quantity of high purity makeup water required for the Tokamak. These include, but are not limited to the designs summarized below:

* Pretreatment, 3 or 4 bed ion-exchange demineralizer, vacuum degasifier, and demineralized water storage

* Pretreatment, reverse osmosis (RO), polishing mixed-bed ion exchange, vacuum degasifier, and demineralized water storage

* Pretreatment, RO, electrodeionization, vacuum degasifier, and demineralized water storage

* Pretreatment, double pass RO, electrodeionization or ion exchange for polishing RO effluent, primary demineralized water storage prior to vacuum degasification (also able to accept water from tanker and to recirculate through polishing demineralizer), polished demineralized water storage in "protected" storage tank (nitrogen blanket, nitrogen sparging, floating roof, or combination of protection)

A major advantage to membrane systems is continuous operation without the need for regeneration and storage of hazardous regenerant chemicals (sodium hydroxide and either sulfuric or hydrochloric acid). Membrane systems are also effective for organics removal. The disadvantages of membrane systems include greater complexity in operation compared to ion exchange systems, requirement for good pretreatment and greater sensitivity to fouling. Membrane systems also prefer continuous rather than intermittent operation. Ion exchange systems better tolerate intermittent operation.

Consequently, membrane systems are often designed with lower flow rates and greater storage capacity compared to all ion exchange systems.

Additional alternatives to an on-site, Project owned and operated makeup demineralizer system include a vendor operated system, a vendor owned and operated system that may be located on-site or just off-site with on-site demineralized water storage and a vendor owned and operated system that also includes demineralized water storage. With vendor supplied demineralized makeup water, the details of demineralizer design may be left to the vendor as long as demineralized water quality and flow requirements are guaranteed. Whichever option is chosen, sufficient on-site demineralized water storage should not be overlooked.

294e Effect of Water Chemistry on Aqueous Corrosion of Copper

Tamara I. Petrova, Water and Fuel Technology Department, Moscow Power Institute (Technical University), 14 Krasnokazarmennaya, Moscow, Russia

Copper-based alloys have been widely used in water-steam cycles in utility industry worldwide and will be used in cooling water system of ITER fusion reactor. Aqueous corrosion and corrosion control of copper alloys present one of the most important issues faced by utility industry. Copper corrosion is governed by multiple factors, of which the most important are:

•Water chemistry.

•Water temperature.

•Flowrate.

•Value of pH and concentration of different impurities.

Presence of copper corrosion products in water results in their deposition on heat transfer surfaces, which decreases heat exchange between metal surface ad water. If water loop is subjected to radiation, corrosion products become activated and increase total coolant activity.

This paper presents test results on the effect of two types of water chemistry – oxidizing and reducing - on copper corrosion rate in water over temperature range 25 – 250 °C. Specific conductivity of water during the tests was <0.15 μS/cm; oxygen or hydrogen was injected into water. Minimum corrosion rate of copper over the studied temperature range was observed during hydrogen injection into water with dissolved hydrogen concentration 30-60 ppb. Surface analysis of specimens showed that at these levels of dissolved hydrogen in water copper was in metallic form. This is in agreement with Pourbaix diagrams developed for this temperature range.

AIChE Annual Meeting
TG005 Fusion System Engineering

Title: Systems Engineering and Risk Management on the National Compact Stellarator Project (NCSX)

Authors: Robert T. Simmons, Philip J. Heitzenroeder, Wayne T. Reiersen, and George H. Neilson (all from PPPL)

Overview

The National Compact Stellarator Experiment (NCSX) was the first of a new class of stellarators known as "compact stellarators." The differentiating feature of a compact stellarator as compared to the traditional stellarator is the use of a "quasi-axisymmetric" magnetic fields to accomplish shaping and confinement. This property permits a more compact device with performance characteristics similar to the well-developed tokamak concept. The advantage of a stellarator is this it is not as prone to disruptions and can be steady state in operation. Currently, there are 13 operating stellarators in the world, ranging from university scale devices to LHD, the world's largest operating fusion experiment One additional stellarator, Wendelstein 7-X, in Germany is under construction. The NCSX Project was managed by PPPL in partnership with the Oak Ridge National Laboratory. The NCSX was a highly developmental project, which distinguished itself from most other DOE construction projects. Unfortunately, primarily due to budget constraints, this project was terminated in May 2008.

This paper will address several innovative approaches and lessons learned in systems engineering that were applied to the NCSX Project:

- Electronic File Systems - because the NCSX Project responsibilities were distributed between PPPL and ORNL, it was necessary to adopt innovative systems engineering approaches to ensure ready transfer and sharing of project information. This was found to be highly successful means of both archiving and providing the necessary ready access to project information.
- Systems Engineering Approaches to Design Requirement Definition – the NCSX Project contained a very detailed and rigorous definition of the requirements for a successful design review and continuous technical oversight of the many work breakdown structure (WBS) elements to assure that the Project's technical requirements would be met. This template for what constituted a "successful" design review proved very useful to ensure that the necessary documentation was prepared.
- Risk Management Systems - because the NCSX Project is a "first-of-a-kind" fusion device that is very complex and has extremely tight installation and fabrication tolerances, it was important to develop a very rigorous approach to defining and managing risks that could threaten the project. It also provides opportunities to improve project cost and schedule performance and the achievement of project technical objectives by identifying and highlighting the risks with the highest potential impacts so that they can be focused on. One of the lessons-learned was that early accurate risk identification and quantification is needed to accurately estimate contingency needs.
- Web Based Meetings – with NCSX being a joint project of PPPL and ORNL, frequent web-based meetings were essential. During times of peak activity, such meetings were held virtually every day. These meetings also proved to be invaluable for communications with our industrial subcontractors, especially when non-conformances needed to be discussed and significantly reduced the need for time-consuming and expensive travel. The effectiveness of

web-based meetings and conference calls is improved when the participants know each other and can accurately interpret the full range of information conveyed in discussions.

Electronic Filing Systems

NCSX was a distributed project with specific design responsibilities assigned to either PPPL or ORNL. Because of the geographical separation, the NCSX Project determined the most efficient methods for ensuring ready access to "baselined" and pending design data and other project management and document guidelines was to adopt a web-based electronic filing system. This enabled project personnel (whether at PPPL, ORNL, or other remote site) to readily access project information – it supported collaborations.

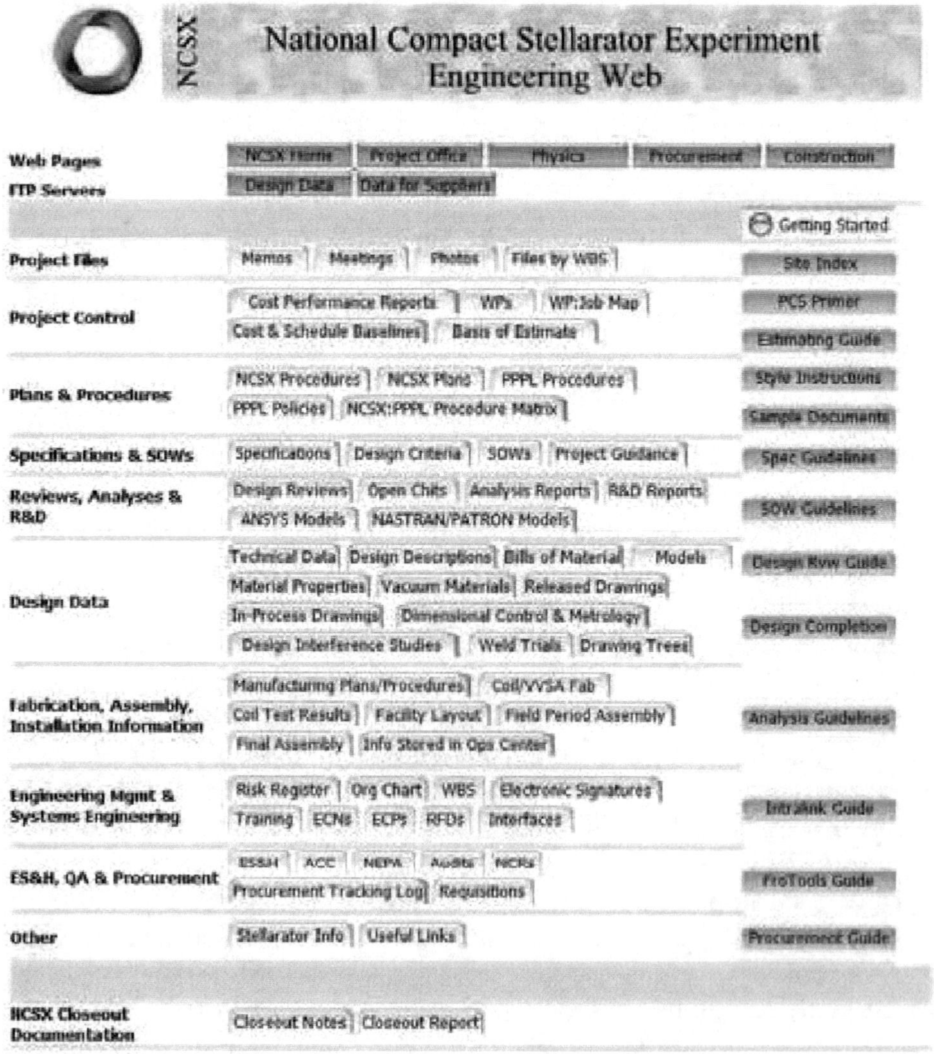

Figure 1 – NCSX Engineering Web Page

The site was maintained on the PPPL web server that was automatically backed up daily. A key feature of this server was that the electronic medium and files were protected by automatic backup software. Proper site security was established to preclude inadvertent altering of documents on the web by restricting "write" access to a limited number of personnel. As necessary, the Project added another level password protection ensure confidentiality of proprietary information

The electronic file system proved to be a very user-friendly and convenient method of data retrieval and exchange for both Project personnel and suppliers At the time of project termination, movement of documents to a particular site for long-term archiving was being performed manually, but plans are being considered on how to possibly streamline and automate this process.

Systems Engineering Approaches to Design Requirement Definition

There were several systems engineering challenges that were addressed by the NCSX Project systems engineering approaches. Most paramount of these was the definition of requirements early in the process. This was achieved at the Project level with a detailed General Requirements Document (GRD) that clearly defined the physics requirements. At the system level, the high-level physics requirements were then translated into the system-level engineering requirements needed to satisfy the GRD requirements. The level of requirements detail was then translated to the individual product or component level and then into the detailed drawings and analyses needed to clearly provide a product that met the NCSX Project requirements.

As with many contemporary engineering projects, the detailed technical data are defined in 3-D CAD models .. These models effectively captured interfaces which obviated the need for much of the more traditional interface documentation previously utilized on earlier projects. For electrical systems, the 2D AutoCAD systems were utilized and configuration control processes followed the common review and approval processes similar to that for written requirements specifications. For mechanical systems, the 3D ProEngineering (ProE) CAD system developed by Parametric Technology Corporation was utilized. In ProE, 3D models are the basic building blocks and these 3D models (global models or parts or assemblies) in conjunction with bills of material and a limited set of drawings whose primary purpose was to convey information required and notes The challenge facing NCSX was how to implement configuration The challenge facing NCSX was how to implement configuration control on 3D models – how does one "sign" 3D models in the normal configuration control processes? Fortunately, ProE (and similar CAD systems such at CATIA that is utilized on ITER) has an integral subsystem (called INTRALINK in ProE) that established the protocols and configuration control processes for models, parts, assemblies, and drawings - changes to models were automatically reflected in changes in drawings and vice versa. INTRALINK had its own set of protocols for controlling changes to models and drawings external to whatever processes implemented by the Project.

It was recognized early on in the Project that a crucial factor in successfully defining the technical, cost, and schedule baselines revolved around designing and implementing an effective revision and configuration process. The Project divided the process into both administrative and technical revision and control processes. For administrative plans and procedures that did not impact technical, cost, or schedule baselines or documentation, the Project maintained a simple revision control process (e.g., signed revisions). A more formal configuration control process was applied to the technical documentation that impacted the technical, cost, or schedule baselines – defined by specifications, analyses, interface control documents, data curves, models and drawings. The Project implemented a rigorous Engineering Change process defined by Engineering Change Proposals (ECPs) that focused primarily on the proposed changes to technical requirements documents, the cost baseline, and the schedule baseline. For drawings, a well-established PPPL Engineering Change Notice (ECN) was utilized to define proposed changes to drawings. These were also controlled electronically. As a means to implement a slightly improved process of configuration control, just prior to a major design review (Conceptual - CDR, Preliminary - PDR. or final - FDR), the NCSX Project team gathered the complete drawing package set and established a "snapshot" of them as of a particular date (usually just before the design review) and established a locked "baseline" in INTRALINK => this "snapshot" contained models and drawings at various levels of promotion (or review an approval) represented the technical baseline against which changes are measured. While this internal project process proved effective for NCSX design reviews, the designers and engineers continued to have the freedom and were encouraged to consider alternate designs as the design evolved until a particular model and associated drawing were promoted to "Release for Fabrication" status at which time more formal configuration control process utilizing the ECN was imposed.

The complexity of the NCSX Project configuration and the tolerance requirements inherent in a three-dimensional structure presented unique challenges. Notwithstanding this, the NCSX Project was indeed a technological success up to the point of termination. NCSX certainly benefitted from several state-of-the-art tools and techniques, such as three- dimensional computer-aided design modeling, finite element analyses, an array of metrology tools which included laser trackers, multi-linked component measuring machines, and photogrammetry systems, and low-distortion welding. Even with extremely tight tolerances, the skill of the technicians and engineers and the use of these tools and techniques enabled the Project to meet its tolerance band. For example, in order to minimize islands in the toroidal flux to less than 10%, a tolerance in the positioning of the modular coil winding pack $\pm < 1.5$ mm was required. Through careful assembly and after-winding shaping techniques the tolerance was achieved on almost all points on the winding path for the modular coils. In the Field Period Assembly process, the three different types of modular coils were aligned, bolted and welded together to form a half period assembly. Alignments were measured to a precision of $\pm$ 0.08 mm and maintained to position requirements of $\pm$ 0.50 mm or less.

The NCSX Project also established a detailed design review process that supplemented existing PPPL requirements. This procedure very clearly defined the requirements for each level of design review (conceptual, preliminary, and final) the specific items, a short definition/clarification of that item, and the success criteria needed to successfully pass that level of design review. This process and "standardized" approach proved very beneficial in clearly defining the necessary levels of documentation and what constituted "success" for each design review.

However, there are some lessons to be learned from the NCSX experience. These lessons include:

- It is essential that adequate R&D and engineering design be developed prior to establishing the cost and schedule baseline. In keeping with this theme, the capabilities of tools and techniques should also be demonstrated and validated before their use is required.
- Future projects should avail themselves of subject matter experts in the design review process. NCSX found that these external experts often brought a fresh and questioning perspective to the design review process.

- o Training and qualification certification standards also need to be included. To improve implementation times and usage estimates, other more experienced users of the technology should be consulted and/or recruited.

Risk Management Processes

Even in the earlier stages of the NCSX Project there was recognition of the importance of risk management in effectively managing a project. Accordingly, risk management techniques were applied to identify risks. The key to successful risk management is alertness to potential risks and the development of a deliberate approach to addressing the risks – either accepting, preventing, mitigating, or avoiding them. The Project Team developed a comprehensive listing of the current known risk item, consequences of the impact of each risk item, and planned or current risk mitigation strategies. These risks were then addressed through design improvements, manufacturing studies, prototypes, schedule contingency, and cost contingency. However, as the design evolved and technical, cost, and schedule performance suffered, it became more evident that a more rigorous approach to risk management was needed. Based on the experience of several other DOE Projects, the project adopted and implemented a risk management process in the spring of 2007, when the Project was approximately 55% complete. Patterned after the concepts called out in the DOE Project Management Order (DOE-O 413.3) and its draft Risk Management Guide (DOE G-413.3-7), a rigorous process of identifying and quantifying all possible risks was started. This process resulted in a very detailed Risk Register listing that documented:

- o A unique identifier assigned to each risk;
- o The impacted jobs that will absorb the risk;
- o A brief description of the risk;
- o The mitigation plan to minimize/eliminate the potential risk;
- o The deadline to retire the risk or to absorb the risk;
- o The owner of the risk;
- o The current status;
- o The likelihood of the risk occurring;
- o The overall consequences of the risk (in broad terms of cost and schedule impacts);
- o The basis of the estimate;
- o The potential cost and schedule (relative to critical path) impacts

Below is a sample of a recent NCSX Risk Register. Since this was done as an Excel spreadsheet, several sorting options were available (e.g., by owner, by risk estimate, etc.). The Risk Register was reviewed every month as part of the formal job statusing process.

NCSX Risk Register

					Last Names only for sorting purposes		Likelihood: VL: P>80% L: 80%>P>40% U: 40%>P>10% VU: P<10% NC: P<1%	Consequences: Negligible <$100k, <0.5 month Marginal: $100k-$500k, 0.5-1 month Significant: $500k-$1M, 1-3 months Critical: $1M-$5M, 3-6 months Crisis: >$5M, > 6 months											730	10517.8
No.	Affected Jobs (absorb the impacts)	Risk Description	Mitigation Plan (& job where budgeted)	Deadline to Retire Risk or Absorb Impact	Owner	Current Status (As of June 2, 2008)	Likelihood of Occurrence	Consequences	Risk Ranking	Basis of Estimate	Cost Impact ($k)	Critical Path Schedule Impact (mos)	Cost and schedule impact calculation basis	consequences			Rank Descriptor	Rank Value	Risk Estimate
2. Assy-02	1815	Station 5: cost and schedule grows when Assembly Sequence Plan fully matures. **Do we update models for Station 6 & 6?**	Expedite Component Designs and Assembly Sequence Plan Jobs 1354, 1501, 1601, 8303	Station 5 & 6 PDR	Brown	**DEFERRED RISK unless we update Station 5 models (end of August completion projected). Then can update and provide costs.**	VL	Significant	High	25% increase in time required for each F.P.	$500	+1.13	*Station 5 estimate of $[illegible] x 20% plus schedule stretchout for the time station 5 is on the c.p. (4.5mo x 25% (stretchout cost quantified and added separately)*	3	3	3	VL3	30	$1,063
2. Assy-03	7504	Station 6: cost and schedule grows when Assembly Sequence Plan fully matures. **Do we update models for Station 6 & 6?**	Expedite Component Designs, Plant Layout, and Assembly Sequence Plan Jobs 1701, 1702, 1803, 8215	Station 5 & 6 PDR	Brown	**DEFERRED RISK unless we update Station 6 models (end of August completion projected). Then can update and provide costs.**	VL	Significant	High	15% increase in time required	$650	+2.18	*Station 6 estimate of $4,317k x 15% plus schedule stretchout for the time station 6 is on the c.p (14.5mo x 15% (stretchout cost quantified and added separately)*	3	3	3	VL3	30	$1,738
2. Assy-04	1810/ 1815 7503	Photogrammetry replaces laser tracker for some operations and saves time and money (Opportunity)	Acquire equipment, develop experience, assess potential. New H/W in place & personnel being trained. 1810 / 1815	After Station 2 MCHP #2.	Dudson / Dudek	Photogrammetry is being incorporated into the assembly process. Evaluate after MCHP #2	L	Opportunity	Opportunity	33% reduction in metrology tasks?	($901)	(3.0)	*50% reduction in metrology crew labor cost =$2,729K x 50%. Reduction of metrology dependent tasks on the c.p. =172 days x .5 = 86 days = 4 months (stretchout cost quantified and added separately)*	#N/A	#N/A	#N/A	#N/A	#N/A	-$1,920
2. Assy-05	1810/ 1815 7503	Assembly delayed due to metrology equipment breakdowns or anomalies	Maintain high availability via maintenance contracts, spares, and trained staff. F&OM Div.	After Station 2 MCHP #2.	Pinner / Dudek	Although minor problems have suffered, has not really manifested itself thus far in assembly of half periods. **Will re-evaluate after MCHP #1 filled over VV.**	L	Significant	Moderate	2 occurrences @ 0.5 month each	$0	+1.00	*(stretchout cost quantified and added separately)*	1	3	3	L3	16	$400
2. Assy-06	1810 1815 7503	General purpose tooling/ lifting equipment (e.g. cranes) not available to support the schedule	Budget lift equipment in FPA Jobs 1810, 1815 - **DONE**	After Station 5?	Dudek	**DEFERRED RISK**	U	Marginal	Low	Up to 2 week impact on FPA and critical path.	$0	+0.50	*(stretchout cost quantified and added separately)*	1	2	2	U2	9	$100
2. Assy-07	1302/1352 1354	Permeability of components outside 3m from machine to test cell walls exceed the permeability limit of mu = 1.2.	Analyze existing conditions to determine if an issue exists. At worst case, may have to utilize more expensive low magnetic materials and/or utilize combination of rebiased trim and PF ring coils to mitigate field errors.	~ June, 2008	Brooks Kalish Chrzanowski	**DEFERRED RISK - work stopped on facility models needed to do this work.**	U	Marginal	Low	Estimated cost of obtaining low magnetic materials.	$200	+0.00		2	1	2	U2	9	$80
3. Stat2-00	**TECHNICAL RISKS - Station 2 Assembly**																	#N/A	#N/A
3. Stat2-02	1810	Station 2: Unacceptable distortion in a field period when welding modular coil shims requiring rework and/ or chair installation.	Perform extensive welding R&D and careful monitoring during welding. Develop suitable weld procedures and train welders to minimize likelihood of unacceptable distortion. Job 1810	After first MCHP	Viola	Weld development continues to improve process. May eliminate TIG cost by moving to flux core wire. Weld prep on shear plates being optimized. **Will re-evaluate after first MCHP.**	U	Significant	Moderate	Cut apart and re-weld two coils back together. Nominally a 2.5-man crew in 12 weeks. Fabrication and installation of chair nominally requires 2.5-man crew and ~3 weeks.	$135	+1.00	*Risk has been reduced by mitigation (design and process changes)*	2	3	3	U3	15	$254
3. Stat2-03	1810	Station 2: Unacceptable distortion in a field period when welding modular coil shims requiring complete disassembly and redesign and reassembly	Perform extensive welding R&D and careful monitoring during welding. Develop suitable weld procedures and train welders to minimize likelihood of unacceptable distortion. Job 1810	After first MCHP	Viola	Weld trials should mitigate probability of occurrence. **Will re-evaluate after first MCHP.**	VU	Critical	Moderate	Cut apart and re-weld two coils back together. Nominally a 2.5-man crew in 12 weeks.	$150	+3.00		2	4	4	VU4	12	$165
3. Stat2-09	1810	Station 2 - shim bag rupture & requires replacement	Pre-qualify shim bags. Use wing chair in compression. Could require taking coil apart in worse case. Job 1810	After last shim bag installed	Viola	**DEFFERED RISK.** Considered unlikely due to other mitigation plan options.	VU	Critical	Moderate	Rough assessment of cost and schedule impact of disassembling/re-assembling coils. [illegible]	$200	+3.00		2	4	4	VU4	12	$175
3. Stat2-10	1810	Station 2. Nose opens up while tightening outboard bolts	Change bolt tightening sequence. Evaluate whether loose pucks are acceptable. Job 1810	After first B1C1 joint successfully completed	Viola	**Will re-evaluate after first B1C1 joint completed.**	U	Marginal	Low	Up to 2 week impact on FPA and critical path	$0	+0.50		1	2	2	U2	9	$100

RR-Rev33.xls Page 1
8/15/2008

Figure 2 – Sample NCSX Risk Register

Early risk mitigation will likely have less cost impact in the early phases of a project as there are more opportunities to address and/or mitigate the risks than in later stages. Later risk assessments may have larger impacts on cost and schedule than earlier risks and the parameters surrounding the later phases may have to be adjusted as required. Certainly implementing a rigorous risk management process late in the design/construction phase of NCSX limited the flexibility to respond. NCSX utilized a Monte Carlo simulation to translate risk into both cost and schedule contingencies. What is evident was the NCSX was one of the first projects within the Office of Science complex to fully utilize this Monte Carlo simulation. Of course, the another potential lesson learned is that the use of such simulations can only be as good as the input data. Effective risk analysis depends on having a good understanding of the design and fabrication issues. Thus, while early risk mitigation is in general preferred, risk analysis must be updated as the design matures to properly capture risks that may be overlooked in the early stages.

Web Based Meetings

The NCSX Project was a distributed project with design responsibilities split between PPPL and ORNL. The use of web-based meetings provided the essential means by which engineers and physicists at offsite locations could easily share and discuss design details, R&D results, and potential solutions to design and installation issues. Design reviews were also frequently conducted with reviewers located elsewhere (nationally or internationally) .This remote meeting capability was also very useful in real-time discussions with subcontractors on discussions of non-conformances and/or design questions. The Project used several commercial web meeting programs (e.g, NetMeeting, MyMeetings, etc.); Each provide unique capabilities, with the trade-off often being ease-of-use versus additional capability

Final Thoughts

The Systems Engineering and Risk Management processes utilized on the NCSX Project were in many instances innovative for projects within the DOE complex. The concepts proved sound, and provided reasonable paths forward to address issues as they arose.. When issues did arise, it was our experience that frequently the implementation details were lacking rather than that the systems engineering and risk management systems and processes were flawed.

Some final thoughts for future projects to consider:

- Complete requisite R&D and designs prior to establishing a baseline. The complex geometry and tight fabrication tolerances of NCSX created unique engineering and assembly challenges. R&D and design needs to be sufficiently completed to establish a sound technical basis for the cost and schedule estimates and risk assessments. To the extent that such tasks are still outstanding at the time a baseline is established, it poses a risk which must be recognized, quantified, and managed with risk acceptance/mitigation/transfer plans and with contingency management.
- Implement rigorous, disciplined cost estimating techniques which factor in the inevitable effects of the learning curve on costs and schedules, especially when a state-of-the art project is being estimated. It is important to realistically assess the uncertainties, their sources, and the prospects for reducing them. Comparison with previous similar experience can be misleading if it does not adequately take into account the special circumstances of a uniqueness and complex of the project. For first-of-a-kind hardware, estimates need to realistically account for "learning experience curves" associated with the initial fabrication, installation, and integration activities.
- The use of formal risk and opportunity assessment techniques, based on a risk register and analysis of the tasks at the job level, is required to establish the need for cost and schedule contingency. It is important to be able to transform the risks identified in the risk registry into contingency requirements, and to help distinguish cost estimation uncertainty from risk. An up-to-date risk registry including risk mitigation actions has to be a key project management tool. As the NCSX Project progressed, the Project Team did become more skilled at recognizing the risks in the remaining work, quantifying them, and developing mitigation plans.
- Complex and geographically distributed projects need to maximize their usage of electronic data base, electronic signing, and web-based meetings to increase the effectiveness of the work process.

INTEGRATED SYSTEMS ENGINEERING APPROACH TO MANAGE THE INSTALLATION AND START UP OF A FUSION REACTOR

S. Chiocchio, I. Kuehn,
ITER-Organization, Cadarache, St Paul Lez Durance F-13108, France

Abstract

ITER is an international research and development project involving the European Union, Japan, the People´s Republic of China, India, the Republic of Korea, the Russian Federation and the USA that aims at demonstrating the scientific and technical feasibility of producing fusion power in magnetically confined plasma. The realization of the project is overseen by the ITER Organization (IO) which was established after the ratification (in October 2007) of the ITER Joint Implementation Agreement signed on the 21 November 2006. The construction activities have started with the preparation of the site, which is located at Cadarache in the South of France, and the procurement of large systems are scheduled to start at the beginning of 2009. ITER will be built and commissioned in the next 10 years and must operate for the following 20 years.
The large number of different engineering disciplines essential for its operation, the worldwide distribution of the design activities and the unusual procurement scheme based on a combination of in-kind and directly funded deliverables makes ITER one of the most challenging projects ever conceived.
The authors have been involved for several years in the definition and practical implementation of the design integration and configuration control structure inside ITER and in the management of system engineering processes.
Here they provide a brief overview of the status of the project and illustrate the systematic approach that the IO has adopted to ensure the consistency of the design with the required performance and describes the management methods, tools and working practices that are being used to facilitate the communication and manage the collaboration among the institutions and industries involved in the project.

1 Introduction

Nuclear fusion occurs when the nuclei of light elements come together at very high temperature and density and fuse together to produce a nucleus of a heavier element, releasing in the process a large amount of energy. This is the process that powers the sun and the stars and the quest to reproduce this phenomenon on the earth initiated soon after this process was first theoretically explained and then experimentally demonstrated in the middle of last century. Different types of devices have been developed and tested. and ITER belongs to the most promising family of these machines, the so called tokamak, initially developed by Russian scientists in the late 50s. In a tokamak fusion reactor a gas formed by a mixture of hydrogen isotopes (deuterium and Tritium) is first ionized (through electrical discharges similar to those in a neon tube) and then heated and compressed using large magnetic fields generated by different sets of coils. The same fields are used to confine and controls the resulting plasma, to prevent that it comes in contact with the physical walls of the device.

The current configuration of ITER is the result of a long design process. The current scientific objectives and main engineering parameters were defined in 1999, when the ITER design team was asked to develop a reduced cost option of a previous design. During this re-design phase the design solutions previously adopted were critically reviewed and areas were identified where simplifications were possible and where the requirements could be relaxed without compromising the overall mission of the project. A key decision during this phase was to exploit most of the design solutions and technologies developed and tested for the large ITER (Barabaschi, Chiocchio, 2003). This permitted an accelerated definition of the machine configuration, but had the downside to make difficult to track the design solutions adopted for each systems to clearly defined requirements.

ITER will be built largely through in-kind contribution by the 7 Parties. The sharing of the procurement scopes have been agreed already in an early stage of the design development and has led to the definition of about 200 Procurement packages with many interfaces among procurements belonging to different Parties. The IO is not directly involved in the management of the industrial contracts, but has the role to define the procurement specifications, that varies from build-to-print specifications to functional specifications and to integrate the whole process during the design and assembly phase. The procurement of such devise through such large international collaboration is an experiment in its own (Chiocchio, 2006). While in many other large industrial enterprises, such as in the nuclear industry, it is not unusual that an industrial main contractor can be selected who has the technical capabilities to deliver the whole plant turn-key, in the case of a fusion machine the number of different specialized engineering disciplines, the experimental nature of the project and the strong interplay between the design choices and the physics performance is such that the function of architect engineer and integrator cannot be entirely outsourced.

At the start of the construction phase the IO, in agreement with the Parties and the worldwide fusion community, has conducted a thorough review of the whole project, starting with the assessment of the validity of the physics objectives and selection of plasma parameters, and continuing with the analysis of safety and licensing issues and the review of the requirements and design choices of the major systems (Holtkamp, 2008). Organisational and management aspects of the project were also re-assessed. At the end of this project review the IO has been asked to re-baseline the project, defining in detail the requirements of the whole projects and of each system, and to establish an updated reference project schedule and cost baseline.

2 An overview of the ITER design

ITER consists of 32 major systems comprising around 10 millions parts with thousands of interfaces among them that must be identified and controlled.
At the core of the plant there is the tokamak, (see Figure 1), that comprises the Vacuum Vessel (VV) , that provides the ultra high vacuum boundary and the first safety barrier of the machine, the in-vessel components, that have to

withstand the large heat and absorb the neutrons produced by the reaction, the superconducting coils used to generate pulsed poloidal fields and steady state toroidal field and to control the plasma position and shape, the thermal shields, that have to reduce the heat load on the cryogenic systems and the cryostat which provides the vacuum conditions required for the operation of the superconducting magnets. Many auxiliary devices, used to provide additional heating, to monitor and control the plasma and to perform the scientific program which is part of the ITER mission, are allocated in large ports of the VV.
Around the tokamak a number of systems are needed to supply the power and maintain the magnets at cryogenic temperature, to cool the plasma facing structure, to fuel the plasma, to create the vacuum and to control the whole plant. A schematic description of the plant configuration is depicted in Figure 2.

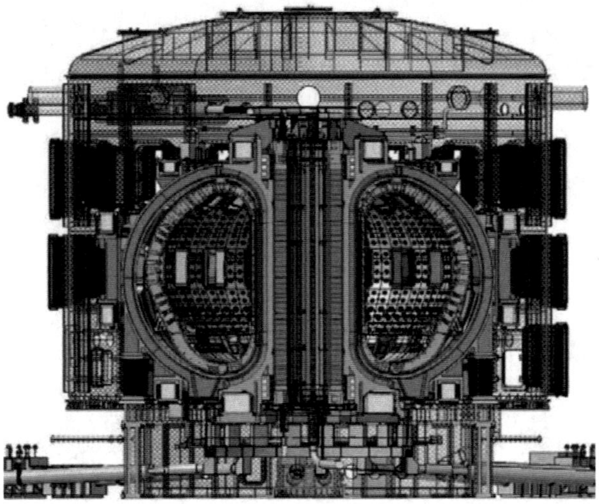

Figure 1 Cross Section through the ITER tokamak

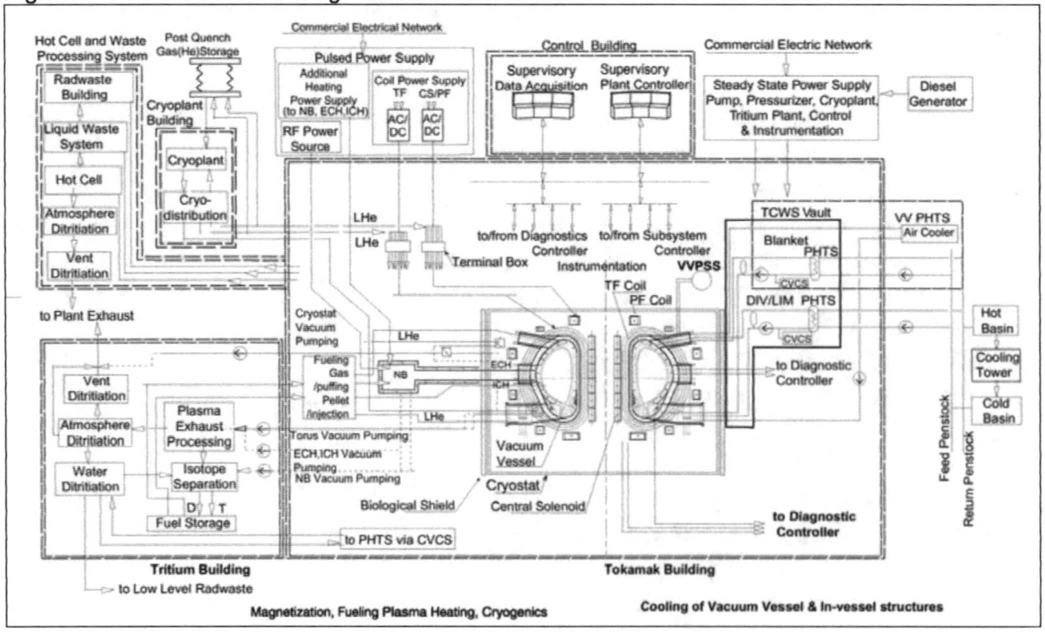

Figure 2. Plant system configuration

The core of the machine and the main auxiliary systems are located in the Tokamak Complex, that includes the Diagnostics, the Tokamak and the Tritium buildings. The 3 buildings are highly integrated and sit on a single concrete slab supported by seismic pads that reduce the effect of horizontal earthquake on the structures inside the buildings. In the inner part of the complex, the tokamak is located in a pit 30 m deep (of which 12 mm below ground) and separated by the rest of the structure by a 2 m thick concrete bio-shield. All around the bioshield, the building is subdivided in 5 levels, that are devoted to specific functions: the lowest level B2 and the highest level (L3) are dedicated to the magnets terminal boxes and to the distribution of the cryogenic fluids, while the three levels corresponding to the three sets of VV ports are allocated to the systems that interface with the plasma. Port Cells (PCs) provide confined space for the execution of remote handling and are separated from the galleries by large steel doors. The galleries provides access to the PCs and are sized to allow the movement of remote operated casks for the transport of activated components and the routing of electric, CODAC and diagnostics lines. Above the tokamak building, the crane hall extends for the whole length and width of the tokamak building and hosts a crane with a total capacity of 1500 tonnes. At the west side of the Tokamak building, the Diagnostic building is hosting diagnostic equipment and control cubicles. The levels B2 and L3 are dedicated for the coil power supply and distribution. At the opposite site the Tritium plant is entirely included inside the Tritium.

The tokamak complex is located approximately at the centre of a large site about 70 hectares that locate all other ITER buildings. (see Figure 3).
The ITER Site is arranged in such a way that all buildings, providing different kind of services to the Tokamak machine, are located around the Tokamak Complex buildings. The site north is hosting the 'Hot Cell Facility' with the 'Rad-waste Building', the 'Personnel Access Control Building' and the area of the 'Neutral Beam Power Supply'. The site west is hosting the electrical facilities providing the site with electrical power and the cryogenic facilities for supply of liquid helium for the coil magnet system. It consist of the 'Pulsed Power High Voltage Substation Area', connecting the grid with the 'Magnet Power Conversion Buildings 32 and 33, and the 'Steady State Power High Voltage Substation Area'.
The site south is hosting the assembly hall for the assembly of the main components of the Tokamak machine. At the east, there are the 'Site Service building', 'Control building and the 'Cooling Water Pumping Station area', which is connecting the Tokamak Complex with the 'Hot and Cold Basins'.

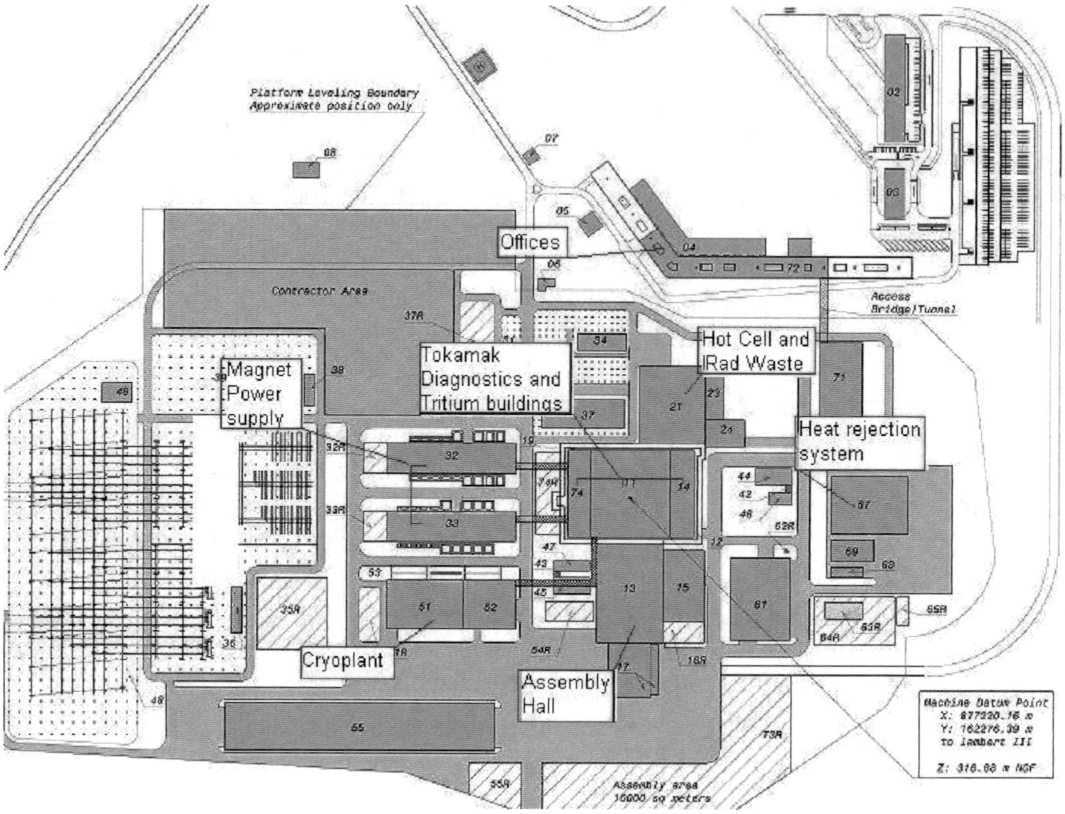

3 Systems Engineering processes in ITER

In early 2007, soon after the signature of the Joint Implementation, we carried out an internal assessment to identify the deficiencies in the design and in the project organization and to define the processes that the IO and its partners should put in place in order to successfully execute the procurement phase of the project.

From the study emerged that while the design of the systems making the core component of ITER, the tokamak, was well advanced, the status of the design of the other systems was very variable. Some systems have been extensively designed, while others were still at a conceptual stage. With such variability in the definition of the design of different interfacing systems, the possibility to effectively integrate the whole plant was questionable and the interfaces with the building structure could not be properly be defined.

Another reason of concern was the large number of design changes being considered as the result of the assessment of the ITER physics basis during the 2007 project review.

The conclusion of the study was that IO had to put higher priority on the finalisation of the Conceptual and Preliminary Design, prior of the commencement of the Detailed

Design for some of the lead procurement systems such as the buildings.

As a result of this assessment, the IO introduced an integrated approach to the development of the design articulated in the following points:

- Establishment of a phased design process covering the whole scope of the design and procurement activities to be carried out using a gated system of design reviews, to assess the maturity of the system design and to validate the consistency of the design with the requirements (Chiocchio et al, 2006);

- Development of a Plant Breakdown Structure, which identify the decomposition of the project in Systems;
- Definition and control the project baseline defining the functional systems requirements and the physical and functional interfaces with other systems;
- Development of a change control process to ensure that the impact of the changes on other systems, on the project performance, cost and schedule is properly assessed and tracked;
- Introduction of an interface management systems that takes into account the progressive definition of the interfaces during the project lifecycle;
- Creation of a centralized design store of the project information accessible to all ITER partners.

ITER design phase definition

Three main stages have been identified during the design phase of the project: conceptual design (CD), preliminary design (PD) and detail design (DD), the latter overlapping with the procurement phase of the project.

During the CD stage, the project top level functional requirements are developed and documented and trade studies are conducted to assess different configuration options. The typical activities during the CD stage include the development of the functional and design requirements, the definition of the design basis and categorization of the loads, the identification of the main interfaces, the first assessment of the cost drivers and the development of a plan to ensure that the reliability, availability, maintainability and inspectability issues are well addressed and that environmental and safety aspects are considered in the design process. At the end of this stage it is expected to have an approved functional specification, schematic of the systems (such as Process Flow Diagrams and top level Process and Instrumentation Diagrams) and the Identification of the systems recorded in Interface Control Documents (ICDs).

During the PD stage, analyses and R&D required to establish the feasibility of the design concepts are completed and verification plans for testing during manufacturing, assembly, and installation are developed. The main parameters at the interfaces with other systems are reviewed and agreed with the interfaces and reported in the ICDs. Typical deliverables include the 3D space allocation models that define the layout of the system, interface drawings, engineering analysis reports to demonstrate compliance with codes and standards, part lists with quantity of items, acceptance criteria, acquisition and installation plans and agreed quality assurance policies.

During the final design or DD stage, the designs of the hardware and software required to satisfy the development specifications are completed. Drawings are released for fabrication. Analyses needed to confirm the adequacy of the detailed designs are performed. R&D required to refine manufacturing processes is also performed. "Build-to" or product specifications are developed and placed under configuration control following a successful formal Final Design Review (FDR).

Following these design activities, the procurement phase include the development of the manufacture drawings and the definition of the technical specifications, the selection of the supplier and the award of the contract, the supervision of the fabrication activities, the on factory acceptance tests the shipping and the on-site acceptance tests. In ITER this part of the lifecycle is normally managed by the Partners in charge of the in-kind procurement, while the IO remains in charge of the integration, configuration control, integration activities on site, include the installation, system commissioning and integrated commissioning and the preparation for the plant operation.

The ITER engineering process and phases are illustrated in Figure 3.

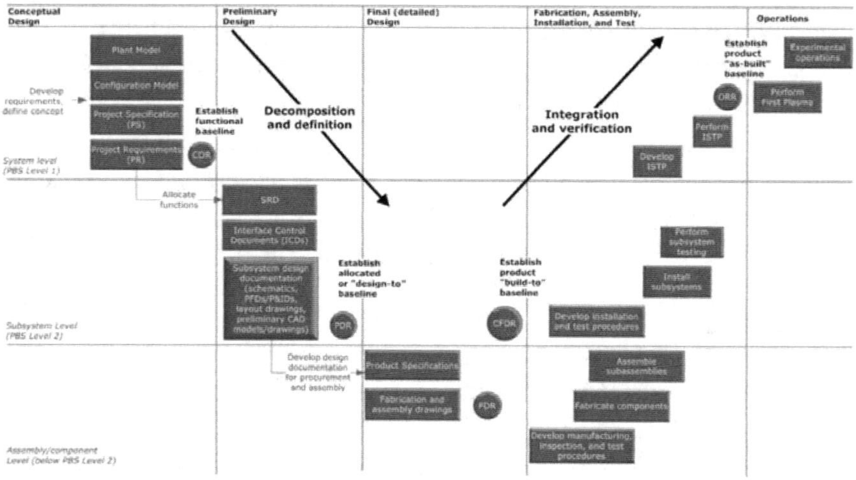

ITER Plant Breakdown structure

The scope of the Plant Breakdown Structure (PBS) is to describe the functional breakdown of the plant in systems, sub-systems, loops, equipment and parts. It ensures that each element of the plant can be uniquely identified as soon it is created during the evolution of the design and manufacturing process, and that all relevant information can be tracked down (or linked) to the corresponding element.

The PBS is organised as a functional tree. Its elements are called "nodes". The nodes are connected in branches and the terminal node of each branch is a "leaf". At each level a node or a leaf correspond to a particular function in the plant.

The leaf of a tree can identify either an elementary part of the plant or an entire assembly. It is expected that the ITER PBS expands with the increasing definition of the design, so that some leaves are replaced by new branches. The ITER PBS consists of 32 major systems which are identified as first level of the PBS.

The ITER Project Baseline

The establishment of a reference baseline that is put under configuration control and can be easily identified and understood by all partners is a key element for the success of a project of the complexity of ITER. Figure 5 illustrates the structure of the ITER baseline. It should be noted that the baselines does not defines only the technical scope of the project, but also establish the costs and resources required to accomplish the task and the schedule to perform and integrate all activities.

The ITER baseline schema also indentifies the approval authority over the elements of the baselines. The top level ITER programmatic objectives (which form the ITER Project Specifications), the Overall Project Cost and Overall Project Schedule are under the responsibility of the ITER council, that represents the ITER stakeholders.at the lower level, the engineering requirements for the whole project, the PBS and WBS, the Integrated Project Schedule and Integrated Project Cost are under the responsibility of the ITER Director General, that has a delegation to approve changes to the baseline that do not impact the overall ITER objectives.

To expedite the management of the changes and non conformance notifications, the baseline defines lower level class of configuration items that are under the responsibility of the technical departments or the Domestic Agencies established by the ITER Parties.

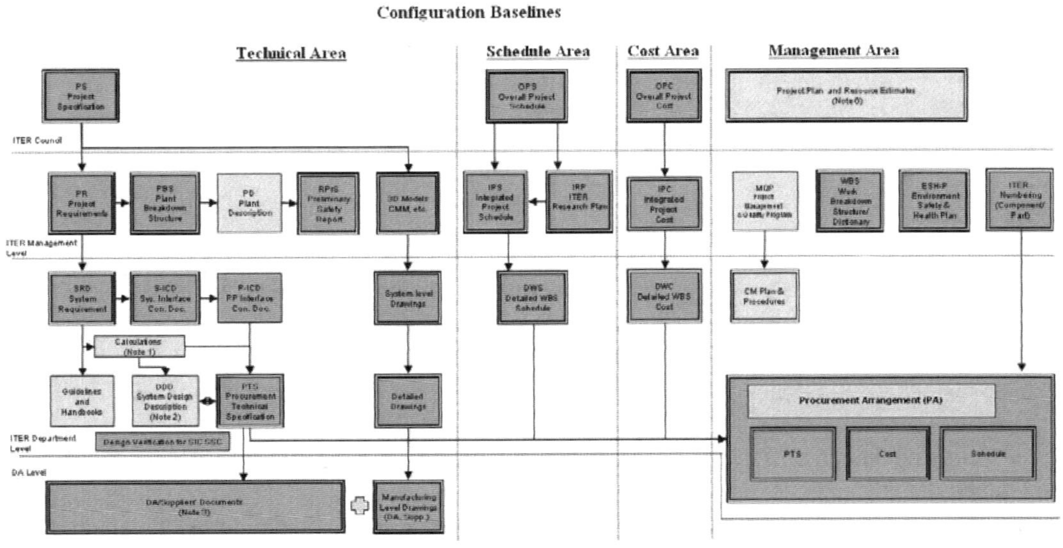

Notes:
0. Documents and drawings in **orange colored boxes** shall be the ITER configuration baselines.
1. Major calculations may be assigned to configuration baselines.
2. DDDs for major systems may be selected as configuration baselines.
3. DA shall assign configuration baseline according to their CM control procedure prepared in accordance with ITER CMP and CI selection criteria.
4. All changes to approved ITER project products shall be processed using PCR.
5. Blue colored boxes do not need CCB and shall be processed with PCR for changes in order to check whether or not affect to cost and schedule aspects.
6. In principle PP&RE is not considered as configuration baseline. However, if PP&RE contains a detailed level of costs assigned to management and department, changes to the cost will be controlled under configuration control.

Figure 5 Structure of the Configuration Baselines

The physical configuration of the plant is represented in a large number of models and drawings that are managed through the ENOVIA LCA (Dassault Systemes, 2008) database. This allows the navigation of the database through geographic or functional based tree structures. It also defines the roles of the different partners and allows the establishment of rules to control the access to the data and the versioning of the models and drawings. In the current phase of ITER, the fast pace of evolution of the design of many systems makes difficult for the engineers accessing the database from remote locations, to identify the current valid configuration. To overcome this difficulty, we have introduced a type of models (Configuration Control Models, CMMs) that define the space envelope and the main interfaces of each systems. These models do not describe the detailed inner structure of the components, and thus they are updated only when modification of the outer boundaries of the component are introduced. The CMMs are also very practical for the detection of clashes and for the graphical representation of large assembly of the ITER plant.

Configuration Management (CM) and Project Changes Procedure.

Two aspects make the management of changes in ITER particular complex. One is the large number of interfaces among the systems; the other, related to the in-kind procurement scheme, derives from the fact that changes to one system under the responsibility to one of the ITER Party often translates in modifications (and extra-cost) that must be taken over by other Parties. The ITER Parties are very sensitive to the need of limiting the number of changes and properly assess their implications.
The ITER change management procedure ensures that all ITER Parties are constantly involved in the assessment of the technical impact of the changes and they participate in the calculation of the related cost and schedule modifications.
The procedure is in place since the creation of the ITER organisation, and to date about 175 Project Change Requests have been processed. Recently we have also introduced a web based system that allows all participants to review the status of all project changes, and be automatically notified when actions are required.

Interface management
The interface management systems in use at ITER is based on the definition of interface control documents (ICD) for each pair of interfacing systems. These documents comprise a scope description, a control table that summarises the status and pending actions for each identified interfaces and a set of Interface Sheets (IS). The IS provides for each interface the:

- *Interface Function*, that specifies “what” the interfacing system must perform;
- *Performance Interface Requirement*, that specifies “how well” a function must be performed (e.g., parameters such as pressure or temperature);
- *Design Interface Constraint*, that specifies applicable codes and standards, specific design requirement for operation and maintenance and essential features;
- *Physical Interface Requirement,* that specifies physical characteristics for components at the interface boundary such as materials, dimensions, tolerances, finish size, weights.;
- *Interface control parameters,* that defines the status of the interface definition (draft, accepted, approved) and the link to the project schedule milestones.

All the ICD and IS are organised in a database that allows the automatic search and control of interfaces. A status report of the interface can be automatically generated by system (listing all interfaces affecting one system) or by pair of interfacing systems (listing of interfaces between them)

Centralised Engineering Database

The IO has already in place an object oriented database called ICP (ITER Collaboration Platform) that covers many of the functionalities of modern Project Lifecycle Management (PLM) tools. The Change Management System and the Interface Management System described above are already implemented in ICP and can be linked to any other "object" defined in the ICP (documents, actions, equipments, people and organizations, etc.).
In parallel the IO is conducting pilot projects to test the functionalities of commercial software that are not implemented in ICP. At present three options are considered, migrate the current functionalities of ICP into a commercial PLM, continue the development of ICP as a full PLM or to integrate the two to reduce the development costs and to cover the entire spectrum of ITER needs.
An example of the ICP functionalities, beside the already mentioned Change management and interface management systems, is the definition of the ITER room database.
The room-book is a centralised tool accessible for everyone working in the ITER Organisation as well as for external contributors via the ITER Technical web page. Its scope is to publish all buildings related information such as concrete outline drawings, general arrangement drawings as well as radiological zoning drawings or magnetic field maps. Each room contains data like volume, wall-, ceiling-, and wall surface, hazards zoning in the different phases and the room environmental conditions such as pressure, temperature or humidity. The room-book also provides the function of a database in terms of retrieving data based on a defined search. Information such as Control Cubicles, magnetic field sensitive components or ferromagnetic objects can be extracted for buildings, levels or rooms of interest.
Different applications, which are configured for the engineering departments, are supported by the room-book such as the 'Steady State Electrical Power Network database' to manage the different electrical loads in the buildings or the 'CODAC database' for identification of the different signals.

4 Conclusions

The management of design activities of a large number of systems to be carried out through a world wide collaboration requires that engineering processes are deployed to control the coherent development of the design and the prompt identification and resolution of issues at the interfaces. Also it is important that efficient and easy to use tools are deployed that improve the communication among all participants.
The ITER organization has recognized the importance of these aspects and has involved the industrial expertise in the definition of systems engineering approach. Also the involvement of all ITER Parties in the definition and assessment of the design processes have been considered a key elements for the success of the international collaboration.

The systematic approach that has been developed and applied and the tools that we have deployed compare favourably with those in use in similar large projects.

5. References

ITER (2001), *Technical Basis for the ITER Final Design*, EDA Documentation Series No 22, IAEA, Vienna 2001.

ITER (2001), *Study of RTO/RC ITER Options,* ITER EDA Document GA0RI199-02-12 W0.2

Barabaschi, P., Chiocchio, S. (2003) *ITER: the Design Basis and the Challenge Ahead*, Proc. of the 20th IEEE/NPSS SOFE Conference, S.Diego;

Chiocchio, S.,(2006) *ITER And International Scientific Collaboration*, Proceedings of EPAC-06, http://accelconf.web.cern.ch/AccelConf/e06/PAPERS/FRYCPA01.PDF

Holtkamp, N., *ITER,* (2008) Proceedings of 25th SOFT Conference, Rostock

Chiocchio, S. *et al.* **(2006)** *System engineering and configuration management in ITER,* Proceedings of 24th SOFT Conference, Warsaw

ENOVIA VPLM V5R16 see http://www.3ds.com

Systems Engineering to Design a Water Cooling System for a Fusion Reactor

Seokho H. Kim, Jan Berry, Juan Ferrada, Kirby Wilcher, Karen McElhaney

US-ITER, Oak Ridge National Laboratory
1055 Commerce Park
Oak Ridge, Tennessee 37830-6483
kims@ornl.gov, (865) 241-1392

US ITER participates in International Thermonuclear Experiment Reactor (ITER) that is constructed in Europe, at Cadarache, France. Specifically, US-ITER is responsible for design, engineering and procurement of the Tokamak Cooling Water System (TCWS). TCWS is designed to provide cooling, baking and chemical control of client systems including first wall, blanket, vacuum vessel, divertor and neutral beam injector. TCWS consists of piping with valves and various fittings, heat exchangers, pumps, pressurizer, etc. Details in system design and engineering procedure, configuration, and various operations will be discussed during the presentation.

The reliability and maintainabilty of the ITER Tokamak Cooling Water System (TCWS) conceptual design were reviewed. The reliability of the TCWS cooling loops was suspect because all elements are required to work in order for the system to support plasma operation. Configuration options for improving the reliability and maintainability of the TCWS cooling loops and chemical and volume control system (CVCS) were proposed.

AUTHOR INDEX